T0155572

SpringerBriefs in Electrical and Computer Engineering

Series Editors

Woon-Seng Gan, School of Electrical and Electronic Engineering, Nanyang Technological University, Singapore, Singapore
C.-C. Jay Kuo, University of Southern California, Los Angeles, CA, USA
Thomas Fang Zheng, Research Institute of Information Technology, Tsinghua University, Beijing, China
Mauro Barni, Department of Information Engineering and Mathematics, University of Siena, Siena, Italy

SpringerBriefs present concise summaries of cutting-edge research and practical applications across a wide spectrum of fields. Featuring compact volumes of 50 to 125 pages, the series covers a range of content from professional to academic. Typical topics might include: timely report of state-of-the art analytical techniques, a bridge between new research results, as published in journal articles, and a contextual literature review, a snapshot of a hot or emerging topic, an in-depth case study or clinical example and a presentation of core concepts that students must understand in order to make independent contributions.

More information about this series at http://www.springer.com/series/10059

Christoph Guger · Natalie Mrachacz-Kersting ·
Brendan Z. Allison

Editors

Brain-Computer Interface Research

A State-of-the-Art Summary 7

 Springer

Editors
Christoph Guger
g.tec Medical Engineering GmbH
Schiedlberg, Austria

Brendan Z. Allison
Department of Cognitive Science
University of California
San Diego, CA, USA

Natalie Mrachacz-Kersting
Department of Health Science
and Technology
Aalborg University
Aalborg, Denmark

ISSN 2191-8112   ISSN 2191-8120   (electronic)
SpringerBriefs in Electrical and Computer Engineering
ISBN 978-3-030-05667-4   ISBN 978-3-030-05668-1   (eBook)
https://doi.org/10.1007/978-3-030-05668-1

This Springer imprint is published by the registered company Springer Nature Switzerland AG
The registered company address is: Gewerbestrasse 11, 6330 Cham, Switzerland

Contents

Brain-Computer Interface Research: A State-of-the-Art Summary 7

Christoph Guger, Brendan Z. Allison and Natalie Mrachacz-Kersting

Abstract Brain-computer interface (BCI) research has been advancing quickly, and novel directions with both invasive and non-invasive BCIs could help new patient groups. Each year, the annual BCI Research Award recognizes the top projects in BCI research. This book includes chapters that review these different BCI projects, and this chapter presents more information about the award process in 2017 and the nominated projects.

Keywords BCI · Brain-computer interface · Award · EEG · ECoG

1 What Is a BCI?

A brain-computer interface (BCI) provides a way for people to convey information without moving. Instead, people perform mental activities, such as imagining movement or counting flashes, that produce specific patterns of brain activity that a computer can detect. The computer then provides feedback in real-time to the user.

All BCIs have four components. First, sensor(s) must detect brain activity. Most BCIs detect brain activity through non-invasive means, usually the EEG, but some BCIs use implanted electrodes that can provide more detailed information about brain activity. Second, automated signal processing software must identify and discriminate brain activity that reflects the user's intent. Third, an external device must provide some sort of feedback based on the outcome of the signal processing module. This could involve presenting a word on a monitor, moving an exoskeleton or

C. Guger (✉)
g.tec Medical Engineering GmbH, Schiedlberg, Austria
e-mail: guger@gtec.at

B. Z. Allison
Cognitive Science Department, University of California at San Diego, La Jolla, USA

N. Mrachacz-Kersting
Department of Health Science and Technology, Aalborg University, Aalborg, Denmark

controlling a prosthetic arm. Finally, an operating environment must control how these other three components interact with each other, and with the end-user.

In the early days of BCI research, most of us were focused on BCIs to provide communication for severely disabled persons. This is still a major goal. However, over the last several years, BCI research has broadened to help patient groups in new ways. For example, BCIs to help persons with stroke regain movement or help patients who require neurosurgery have both been gaining attention in recent years. However, concerns have been raised about the risks of excessive hype and unrealistic statements, which could lead people to distrust BCI research in general and even lead to unethical "false hope" with patients. Non-profit entities like the BCI Society could help encourage effective and appropriate use of BCI technology.

2   The Annual BCI-Research Award

One of the companies known for producing hardware, software, and systems for BCIs, called G.TEC, started the Annual BCI Research Award in 2010. The award was meant to recognize and encourage top-quality BCI research, and provide a way to explore new trends and issues developing in our research field. The award is open to any group worldwide, regardless of which company's tools are used.

In 2017, G.TEC decided to turn the award management to a non-profit entity in Austria called the BCI Award Foundation, chaired by Drs. Christoph Guger and Gerwin Schalk. However, the awards still follow the same procedure and use the same criteria:

- A Chairperson of the Jury is chosen from a well-known BCI research institute.
- This Chairperson forms a jury of top BCI researchers who can judge the Award submissions.
- We post information about the BCI Award for that year, including submission instructions, scoring criteria, and a deadline.
- The jury reviews the submissions and scores each one, based on the scoring criteria below. The jury then determines the nominees and winners. We have expanded from ten to twelve nominees and added second and third place winners since 2010.
- The nominees are announced online, asked to contribute a chapter to this annual book series, and invited to a Award Ceremony that is attached to a major conference (such as an International BCI Meeting or Conference).
- At this Award Ceremony, the twelve nominees each receive a certificate, and the winner is announced. The winner earns $3000 USD and the prestigious trophy. The 2nd place winner gets $2000 USD and the 3rd place gets $1000 USD.

The 2017 jury was:
Natalie Mrachacz-Kersting (chair of the jury 2017),
Gaurav Sharma (Winner 2016),
Reinhold Scherer,
Jose Pons,

Femke Nijboer,
Kenji Kansaku,
Jing Jin.

Consistent with tradition, the jury included the winner from the preceding year (Gaurav Sharma). The international jury included BCI experts working in Japan, China, the US, and four European countries. The chair of the jury, Dr. Natalie Mrachacz-Kersting, is a top figure in BCI research and leads the prestigious BCI lab at Aalborg University, Denmark. Dr. Natalie Mrachacz-Kersting said: "I was very fortunate to work with the 2017 jury. All of the jury members that I approached chose to join the jury, and we had an outstanding team!"

How does the jury decide the nominees and winners? We have used the same scoring criteria across different years. These are the criteria that each jury uses to score the submissions. Earning a nomination (let alone an award) is very challenging, given the number of submissions and the very high quality of many of them. Submissions need to score well on several of these criteria:

- Does the project include a novel application of the BCI?
- Is there any new methodological approach used compared to earlier projects?
- Is there any new benefit for potential users of a BCI?
- Is there any improvement in terms of speed of the system (e.g. bit/min)?
- Is there any improvement in terms of accuracy of the system?

Fig. 1 The picture shows the head of the jury, Natalie Mrachacz-Kersting, reading the nominee certificates at the BCI Award ceremony at the BCI conference in Graz 2017

Fig. 2 Christoph Guger (organizer), Brendan Z. Allison (moderator), Natalie Mrachaz-Kersting (head of jury), Doron Friedman (nominee), Femke Nijboer (member of jury), Reinhold Scherer (member of jury)

- Does the project include any results obtained from real patients or other potential users?
- Is the used approach working online/in real-time?
- Is there any improvement in terms of usability?
- Does the project include any novel hardware or software developments?

After totaling the scores for all projects, the next step in the award process is the awards ceremony. Each year, we host a Gala Awards Banquet that is attached to a major conference, where we review the nominees and announce the winners. We also ask nominees to come onstage and receive a certificate and other prizes—though the biggest prize is recognition in front of hundreds of peers. All of the winners also receive a cash prize, and the first place winner earns a prestigious trophy (Figs. 1 and 2).

3 The BCI Research Award Book Series

We published the first BCI Research Award book to recognize the projects that were nominated for the first BCI Research Award in 2010. Since then, we have continued to produce a book each year that presents the projects from that year's competition. In addition to presenting the work that was nominated, the chapters also include follow-up work, discussion, future issues, and other new material.

Each book starts when the nominees are announced. We then ask the nominees if they can provide a chapter for us. We usually establish a deadline several months after the award ceremony. The editors then review the chapters, then send the authors comments for improvement. In some cases, only minor revisions are needed; other times, we may ask for revised figures or substantial detail. We then send these chapters to the publisher, which sends back proofs that we review.

We have also included an introduction and discussion chapter in each book. In the discussion, we review trends from the annual awards. Notably, many of the prominent emerging directions in BCI research that we mentioned above—such as BCIs for stroke rehabilitation or neurosurgery—are also addressed in the forthcoming chapters. The chapters also present fairly unexplored goals for patients, such as reducing phantom limb pain or adapting to lapses in attention during therapy. This book will thus introduce readers to some of the newest and most promising projects in BCI research, which could spearhead new systems and methods to help patients in new ways. The chapters from this year's awards also address fundamental scientific issues such as how we produce speech or movement.

4 Projects Nominated for the BCI Award 2017

This year's jury reviewed all of the submissions based on the scoring criteria presented above. After tallying the scores across all reviewers, the twelve submissions that were nominated for a BCI Award 2017 were:

Takufumi Yanagisawa[1–6]*, Ryohei Fukuma[1,3,4,7], Ben Seymour[8,9], Koichi Hosomi[1,10], Haruhiko Kishima[1], Takeshi Shimizu[1,10], Hiroshi Yokoi[11], Masayuki Hirata[1,4], Toshiki Yoshimine[1,4], Yukiyasu Kamitani[3,7,12], Youichi Saitoh[1,10]

BCI prosthetic hand to control phantom limb pain

1. Osaka University Graduate School of Medicine, Osaka, Japan.
2. Osaka University Graduate School of Medicine, Osaka, Japan.
3. ATR Computational Neuroscience Laboratories, Kyoto, Japan.
4. CiNet Computational Neuroscience Laboratories, Osaka, Japan.
5. JST PRESTO, Osaka, Japan.
6. Osaka University, Japan.
7. Nara Institute of Science and Technology, Nara, Japan.
8. University of Cambridge, UK.
9. National Institute for Information and Communications Technology, Osaka, Japan.

10. Osaka University Graduate School of Medicine, Osaka, Japan.
11. The University of Electro-Communications, Tokyo, Japan.
12. Kyoto University, Japan.

Mariska J Vansteensel[1], Elmar GM Pels[1], Martin G. Bleichner[2], Mariana P. Branco[1], Timothy Denison[3], Zachary V. Freudenburg[1], Peter Gosselaar[1], Sacha Leinders[1], Thomas H. Ottens[4], Max A Van Den Boom[1], Peter C. Van Rijen[1], Erik J. Aarnoutse[1], Nick F Ramsey[1]

Implantable Communication Brain-Computer Interface for Home-Use in Locked-In Syndrome

1. Brain Center Rudolf Magnus, University Medical Center Utrecht, Dept Neurology and Neurosurgery, Heidelberglaan 100, 3584 CX Utrecht, The Netherlands.
2. Neuropsychology Lab, Department of Psychology, European Medical School, Cluster of Excellence Hearing4all, University of Oldenburg, Ammerländer Heerstrasse 114-118, 26129 Oldenburg, Germany.
3. Neuromodulation Core Technology, 7000 Central Ave NE, Medtronic PLC, Minneapolis, 55432, USA.
4. University Medical Center Utrecht, Dept Anesthesiology, Heidelberglaan 100, 3584 CX Utrecht, The Netherlands.

Tomislav Milekovic[1], Marco Capogrosso[1,2], David Borton[1,3], Fabien Wagner[1,£], Eduardo Martin Moraud[2,£], Jean-Baptiste Mignardot[1], Nicolas Buse[4], Jerome Gandar[1], Quentin Barraud[1], David Xing[3], Elodie Rey[1], Simone Duis[1], Yang Jianzhong[5], Wai Kin D. Ko[5], Qin Li[5,6], Peter Detemple[6], Tim Denison[4], Silvestro Micera[2,8,&], Erwan Bezard[5,6,9,10,&], Jocelyne Bloch[11,&], Grégoire Courtine[1,11]

A brain-spine interface to alleviate gait deficits after spinal cord injury

1. International foundation for Research in Paraplegia chair in Spinal Cord Repair, Center for Neuroprosthetics and Brain Mind Institute, School of Life Sciences, EPFL, Switzerland.
2. Bertarelli Foundation Chair in Translational Neuroengineering, Center for Neuroprosthetics and Institute of Bioengineering, School of Bioengineering, EPFL, Switzerland.
3. School of Engineering, Brown University, USA.
4. Medtronic, USA.
5. Motac neuroscience Ltd., UK.
6. Institute of Lab Animal Sciences, China Academy of Medical Sciences, China.
7. Fraunhofer ICT-IMM, Germany.
8. The BioRobotics Institute, Scuola Superiore Sant'Anna, Italy.
9. University of Bordeaux, Institut des Maladies Neurodégénératives, UMR 5293, France.
10. CNRS, Institut des Maladies Neurodégénératives, UMR 5293, France; 11 Centre Hospitalier Universitaire Vaudois, Switzerland.

Ujwal Chaudhary[1,2], Aygul Rana[1], Azim Malekshahi[1], Stefano Silvoni[3], Niels Birbaumer[1,2]

Brain Computer Interface for Communication with Patients in Completely Locked-in State

1. Institute of Medical Psychology and Behavioral Neurobiology, University of Tuebingen.
2. Wyss Center for Bio and Neuroengineering, Genéva, Switzerland.
3. Department of Cognitive and Clinical Neuroscience, Central Institute of Mental Health, Mannheim, Germany.

Stephanie Martin[1,2], Peter Brunner[3,4], Iñaki Iturrate[1], José del R. Millán[1], Gerwin Schalk[3,4], Robert T. Knight[2,5] & Brian N. Pasley

Individual word classification during imagined speech

1. Defitech Chair in Brain-Machine Interface, Center for Neuroprosthetics, Ecole Polytechnique Fédérale de Lausanne, Switzerland
2. Helen Wills Neuroscience Institute, University of California, Berkeley, CA, USA
3. National Center for Adaptive Neurotechnologies, Wadsworth Center, New York State Department of Health, Albany, NY, USA
4. Department of Neurology, Albany Medical College, Albany, NY, USA
5. Department of Psychology, University of California, Berkeley, CA, USA

Surjo R. Soekadar, Matthias Witkowski, Cristina Gómez, Eloy Opisso, Josep Medina, Mario Cortese, Marco Cempini, Maria Chiara Carrozza, Leonardo G. Cohen, Niels Birbaumer, Nicola Vitiello

Restoration of finger movements in everyday life environments using a hybrid brain/neural hand exoskeleton

University Hospital of Tübingen, Germany.

David Hübner[*1], Pieter-Jan Kindermans[*2], Thibault Verhoeven[3], Klaus-Robert Müller[2,4,5], Michael Tangermann[1]

Rethinking BCI Paradigm and Machine Learning Algorithm as a Symbiosis: Zero Calibration, Guaranteed Convergence and High Decoding Performance

1. Albert-Ludwigs Universität Freiburg, Germany.
2. Berlin Institute of Technology, Germany.
3. Ghent University, Belgium.
4. Korea University, Seoul, Korea.
5. Max Planck Institute for Informatics, Saarbrücken, Germany.

*These authors contributed equally.

Takasaki K[1], Liu F[2], Hiramato M[2], Okuyama K[2], Kawakami M[2], Mizuno K[2], Kasuga S[1,3], Noda T[4], Morimoto J[4], Fujiwara T[5], Ushiba J[1,3,*], Liu M[2]

Targeted up-conditioning of contralesional corticospinal pathways promotes motor recovery in poststroke patients with severe chronic hemiplegia

1. Department of Biosciences and Informatics, Faculty of Science and Technology, Keio University, Japan.
2. Department of Rehabilitation Medicine, Keio University School of Medicine, Japan.
3. Keio Institute of Pure and Applied Science (KiPAS), Japan.
4. ATR Computational Neuroscience Labs, Japan.
5. Department of Rehabilitation Medicine, Juntendo University School of Medicine, Japan.

*Corresponding author.

Korostenskaja, M.[1,2,3], RaviPrakash, H.[4], Bagci, U.[4], Lee, K.H.[3], Chen, P.C.[1,3], Salinas, C.[3], Baumgartner, J.[3], Castillo, E.[2,3]

Gold Standard for epilepsy/tumor surgery coupled with deep learning offers independence to a promising functional mapping modality

1. Functional Brain Mapping and Brain Computer Interface Lab, Florida Hospital for Children, Orlando, FL, USA.
2. MEG Lab, Florida Hospital for Children, Orlando, FL, USA.
3. Florida Epilepsy Center, Florida Hospital, Orlando, FL, USA.
4. Center for Research in Computer Vision, University of Central Florida, Orlando, FL, USA.

Ori Cohen[1,2], Dana Doron[3], Moshe Koppel[3], Rafael Malach[4], Doron Friedman[1]

High Performance BCI in Controlling an Avatar Using the Missing Hand Representation in Long Term Amputees

1. The Advanced Reality Lab, Interdisciplinary Center Herzliya (IDC H.), P.O. Box 167 Herzliya, 46150, Israel.
2. The Department of Computer Science, Bar-Ilan University Ramat-Gan, 52900 Israel.
3. The Department of Brain Injury Rehabilitation, Sheba Medical Center, Tel-Hashomer, Israel.
4. The Department of Neurobiology, Weizmann Institute of Science, Rehovot 76100, Israel.

S. Aliakbaryhosseinabadi[1], E. N. Kamavuako[1], N. Jiang[2], D. Farina[3], N. Mrachacz-Kersting[1]

Online adaptive brain-computer interface with attention variations

1. Center for Sensory-Motor Interaction, Department of Health Science and Technology, Aalborg University, DK-9220 Aalborg, Denmark.

2. Department of Systems Design Engineering, Faculty of Engineering, University of Waterloo, Waterloo, Canada.
3. Imperial College London, London, UK.

Birgit Nierula[1,2], Maria V. Sanchez-Vives[1,2,3,4]

Which BCI paradigm is better to induce agency or sense of control over movements?

1. Institut d'Investigacions Biomè diques August Pi I S unyer (IDIBAPS), Rosselló 149–153, 08036 Barcelona, Spain.
2. Event-Lab, Department of Clinical Psychology and Psychobiology, Universitat de Barcelona, Passeig de la Vall d'Hebron 171, 08035 Barcelona, Spain.
3. Institució Catalana Recerca i Estudis Avançats (IC R E A), Passeig Lluís Companys 23, 08010 Barcelona, Spain.
4. Departamento de Psicología Básica, Universitat de Barcelona, Passeig de la Valld'Hebron 171, 08035 Barcelona, Spain.

5 Summary

BCIs are increasingly powerful tools that could help new patient groups in new ways. Translating laboratory results to long-term, independent home use that provides real benefits over current treatments requires extensive work. However, the projects nominated for the 2017 BCI Research Award present innovative directions and ideas that could someday lead to next-generation treatments. In addition to presenting chapters reviewing these different projects and related work, this book also presents an analysis of the trends reflected in the awards within our discussion chapter. We hope these chapters provide an enjoyable and informative overview of some of the most promising projects in our field.

Gold Standard for Epilepsy/Tumor Surgery Coupled with Deep Learning Offers Independence to a Promising Functional Mapping Modality

M. Korostenskaja, H. Raviprakash, U. Bagci, K. H. Lee, P. C. Chen, C. Kapeller, C. Salinas, M. Westerveld, A. Ralescu, J. Xiang, J. Baumgartner, M. Elsayed and E. Castillo

Abstract RATIONALE: Electrocorticography-based functional language mapping (ECoG-FLM) utilizes an ECoG signal paired with simultaneous language task presentation to create functional maps of the eloquent language cortex in patients selected for resective epilepsy or tumor surgery. At present, the concordance of functional maps derived by ECoG-FLM and electrical cortical stimulation mapping (ESM) remains rather low. This impedes the transition of ECoG-FLM into an independent functional mapping modality. As ESM is considered the gold standard of functional mapping, we aimed to use it in combination with machine learning (ML) approaches ("ESM-ML guide"), to improve the accuracy of ECoG-FLM. METHODS: The ECoG data was collected from 6 patients (29.67 $\pm$ 12.5 yrs; 19–52 yrs; 3 males, 3 females). Patient ECoG activity was recorded (g.USBamp, g.tec, Austria) during administration of language tasks. For data analysis: (1) All ECoG sites were

---

M. Korostenskaja (✉) · P. C. Chen · M. Elsayed
Functional Brain Mapping and Brain Computer Interface Lab, Orlando, FL, USA
e-mail: milena.korostenskaja@gmail.com

M. Korostenskaja · E. Castillo
MEG Center, Florida Hospital for Children, Orlando, FL, USA

M. Korostenskaja · K. H. Lee · P. C. Chen · C. Salinas · M. Westerveld · J. Baumgartner · E. Castillo
Florida Epilepsy Center, Florida Hospital, Orlando, FL, USA

H. Raviprakash · U. Bagci
Center for Research in Computer Vision, University of Central Florida, Orlando, FL, USA

C. Kapeller
g.tec Medical Engineering GmbH, Schiedlberg, Austria

A. Ralescu
EECS Department, University of Cincinnati, Cincinnati, OH, USA

J. Xiang
Xiang Research Lab, Division of Neurology, Cincinnati Children's Hospital Medical Center, Cincinnati, USA

M. Elsayed
Harvard Medical School, Beth Israel Deaconess Medical Center, Boston, MA, USA

C. Guger et al. (eds.), *Brain-Computer Interface Research*,
SpringerBriefs in Electrical and Computer Engineering,
https://doi.org/10.1007/978-3-030-05668-1_2

divided into ESM positive [ESM(+)] and ESM negative [ESM(−)]; (2) Features of ESM(+) and ESM(−) sites in the ECoG signal were determined by analyzing the signal in the frequency domain; (3) ML classifiers [Random Forest (RF) and Deep Learning (DL)] were trained to identify these features in language-related ECoG activity; (4) The accuracy of the ESM-ML guided classification was compared with the accuracy of the conventional ECoG-FLM. RESULTS: The conventional approach demonstrated: 58% accuracy, 22% sensitivity, and 78% specificity. The "ESM-ML guide" approach with RF classifier demonstrated: 76.2% accuracy, 73.6% sensitivity and 78.78% specificity. The DL classifier achieved the highest performances compared to all others with 83% accuracy, 84% sensitivity and 83% specificity. CONCLUSION: ECoG-FLM accuracy can be improved by using an "ESM-ML guide", making the use of ECoG-FLM feasible as a stand-alone methodology. The long-term goal is to create a tool-box with "ready to use an ESM-ML guide" algorithm trained to provide high accuracy ECoG-FLM results by classifying between ESM(+) and ESM(−) contacts in prospective sets of language-related ECoG data and, thus, contribute towards improved surgical outcomes.

Keywords Brain computer interface (BCI) · Deep learning (DL) · Electrocorticography (ECoG) · Epilepsy surgery · Functional brain mapping · High-gamma mapping · Machine learning (ML) · Passive mapping · Language mapping

1 Introduction

Electrocorticography-based functional language mapping (ECoG-FLM) [1, 2] is rapidly gaining popularity as an approach for localization of eloquent language cortex in patients selected for resective epilepsy surgery [3–6] or tumor resection [7]. Unlike other approaches, ECoG-FLM provides unprecedented flexibility and offers advantages from time, effort and safety perspectives. ECoG-FLM uses the ECoG signal recorded from grid and depth electrodes implanted for clinical purposes as a part of pre-surgical evaluation process (Fig. 1). It allows fast, real-time mapping of the eloquent language cortex at the patient's bed-side for infants (for an example, during a spontaneous cooing- and babbling [8]), for both children and adults (for an example, during a spontaneous conversation [9, 10]), as well as intra-operatively [11, 12].

Language-related ECoG changes are mainly reported in the gamma frequency band [13], leading to its utilization in all ECoG-based functional language mapping batteries [4, 5, 14]. However, the task-related power increase observed in the gamma frequency of the analyzed ECoG-FLM data is not limited to language processing. Indeed, it is rather universal and seen in other modalities, such as motor [15] and sensory [16]. Despite this commonality between the observed task-related gamma frequency power changes, the accuracy of the functional mapping in these modalities is high for sensory and motor, but relatively low for language modality. On

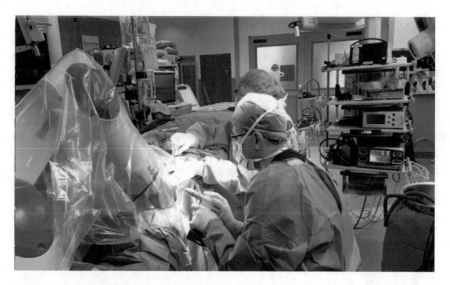

Fig. 1 Depth electrode implantation with Robotized Stereotactic Assistant (ROSA) at Florida Epilepsy Center, Florida Hospital, Orlando, Florida, USA

average, ECoG-FLM had lower sensitivity (62%) and higher specificity (75%) to detect language-specific regions [for a comprehensive review, see 17], which is the opposite than observed for hand motor (100% sensitivity and 79.7% specificity) and hand sensory (100% sensitivity and 73.87% specificity) ECoG-based mapping [4]. Low accuracy hinders the ECoG-FLM transition to become an independent modality utilized routinely in clinical settings.

In order to avoid post-surgical language deficits, language cortex boundaries need to be determined by identifying *both cortex that is essential for language function* as well as *cortex that is nonessential* [18]. To ensure that these substrates are mapped during pre-surgical evaluation, the current gold standard procedure (ESM) is used to guide surgical decision making and resection boundaries [19]. When aiming to develop alternative ECoG-based testing methodology comparable in accuracy to ESM, a mechanism is needed to reliably predict whether each tested ECoG electrode is located over the language or non-language area, as defined by ESM, across multiple subjects. In such situations, where traditional statistical approaches are difficult to apply, the use of machine learning (ML) becomes an invaluable asset [20]. ML has shown great promise in the analysis of complex EEG [21] and multi-dimensional ECoG patterns for brain-computer interfaces (BCIs) [22–25]. Moreover, we have demonstrated that novel ML algorithms can be developed and effectively applied to neurophysiological data analysis in patients with drug-resistant epilepsy [26].

Considering the scientific premise of ML and the "ground truth" ESM value for functional mapping, we combined these two methodologies to "guide" ECoG-FLM towards increased sensitivity, specificity and overall accuracy. We will refer to this approach here as to the "*ESM-ML guide.*" Expected outcome of this innovative approach will be an application of ECoG-FLM as a stand-alone methodology, leading to overall paradigm-shifting changes in clinical use of functional mapping. We aimed to innovate ECoG-FLM signal processing by investigating two advanced classifiers (*Random Forest - RM* [6] *and Deep Learning - DL*) and adapting them to work with ECoG signals. Our latest addition for accurate signal prediction was an introduction of DL algorithms to represent the ECoG signals in a hierarchical and the most compact way. To the best of our knowledge, it is the first time that DL is being used in ECoG-FLM experiments. In this chapter, we summarize our work nominated for the Annual International BCI Award [27] and presented at the 2017 BCI Research Award ceremony during the 7th Graz Brain Computer Interface Conference 2017 in Graz, Austria [28], at American Epilepsy Society Annual Meeting 2017 in Washington, DC, USA [29] and at the IEEE International Conference on Systems, Man, and Cybernetics (SMC) 2017 in Banff, Canada (Best Four Conference Papers Award [30]).

2   Methods

2.1   Study Participants

The ECoG data was collected from six patients (29.67 $\pm$ 12.5 yrs; 19–52 yrs; 3 males, 3 females) with diagnosis of pharmacoresistant epilepsy (Table 1), undergoing comprehensive evaluation for epilepsy surgery at Florida Epilepsy Center, Florida Hospital, Orlando, USA. Their evaluation required extra-operative intracranial electroencephalography (EEG) monitoring using subdural electrodes for further localization of seizure onset zone and for functional mapping. The Institutional Review Board of Florida Hospital approved the study protocol.

2.2   MRI, CT, and ECoG Co-registration

A pre-operative 3D volumetric MRI scan, using a 3T GE scanner (T1 FSPGR, 512 $\times$ 512 matrix, FOV 250, 1 mm slice), and a high resolution post-operative CT scan (512 $\times$ 512 matrix, FOV 250, 1-mm slice thickness) were acquired to visualize the location of subdural electrodes on the brain surface. The MRI and CT scans were co-registered using a free-form deformation-based non-linear registration algorithm (commercial software AnalyzeDirect, Inc. Overalnd Park, KS). Additionally, 3D segmentation of the brain surface and grid positions were performed using the same platform (Fig. 2). Both registration and segmentation results were evaluated and validated by participating experts to make sure the results were clinically feasible.

Table 1 Patient demographics (N = 6). All patients were adults (age 18 and older), left hemisphere dominant for language, with left or bilateral grid coverage

Pt#	Age, yrs	Sex	DH	HLD	Epilepsy focus	Side of focus	Grid placement	Epilepsy onset, yrs	Epil duration, yrs	VCI	WMI	PSI	FSIQ	PIQ	VIQ
1	19	M	R	L^W	F-T	L	L	16	3	81	80	111	95	111	81
2	33	F	R	L^W	F, T	L	L	10	23	99	95	84	102	104	99
3	20	M	R	L^{MEG}	F-T	L	L	6	14	102	80	62	100	97	102
4	22	F	R	L^W	P	L	L	20	2	74	95	74	77	85	74
5	32	F	R	L^W	T	L	B	26	6	87	86	76	78	74	87
6	52	M	R	L^W	T	L	L	30	22	100	95	94	107	115	100

F female, M male; DH dominant hand, R right, L left, B bilateral; HLD hemispheric language dominance, L^W left language dominance determined with WADA procedure, L^{MEG} left language dominance determined with magnetoencephalography (MEG); F frontal, T temporal, P parietal; VCI Verbal Comprehension Index, WMI Working Memory Index, PSI Processing Speed Index, FSIQ Full Scale IQ, PIQ Performance IQ, VIQ Verbal IQ

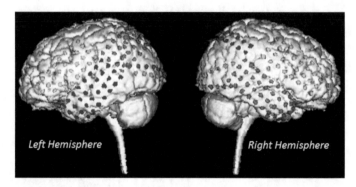

Fig. 2 Reconstructed 3D-model of the brain from pre-operative MRI scan co-registered with a post-op CT scan to transfer electrode location information into MRI scans for localization of structure and function of interest. The dots represent a *bilateral hemispheric coverage* with ECoG electrodes (strips and grids) in study participant #05. The coverage extended to both left and right frontal, temporal, parietal and occipital lobes

2.3 Data Recording

Patients' baseline ECoG activity for localizing the eloquent language cortex was recorded with the subdural grid electrodes, implanted for clinical purposes. The clinical ECoG data was recorded via a clinical EEG system (Nihon Kohden, Irvine, CA). A splitter box was used to simultaneously acquire the research data at a sampling frequency of 1200 Hz using the g.USBamp amplifiers (g.tec, Austria). This approach ensured that ECoG recording for seizure localization remained uninterrupted during the research study procedure [4, 5, 14]. The SIGFRIED toolbox, based on the BCI2000 brain computer interface (BCI) software platform, was used for signal acquisition and visualization of a real-time task-related activity [2]. The following paradigms were used after six minutes of recording the baseline activity, during which the patient was instructed to relax and remain as still as possible: for receptive language function—story processing, reading comprehension; for expressive language function—verb generation, picture naming, counting and alphabet recitation tasks (Fig. 3).

2.4 Data Analysis

First, the ECoG-FLM data was reviewed for artifacts and noisy channels in BCI2000 Viewer software [32]. Afterwards, it was analyzed in two ways: (1) conventional ECoG-FLM analysis [33]; and (2) our new ESM-FLM-guided approach. Yielded results were then compared in terms of accuracy, specificity and sensitivity.

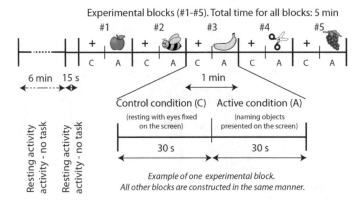

Fig. 3 Experimental set-up of ECoG-FLM: Example of Picture Naming task (also referred to as Visual Naming task) to map expressive language function, (localize Broca's area) (adopted from Ref. [31])

2.4.1 Conventional Approach

All areas of task-related activity were analyzed, with specific attention given to significant activity in the language-associated regions including Broca's and Wernicke's areas due to their known association with speech production and language comprehension, respectively. The data was filtered with notch filters at all harmonics hi of the 60 Hz power line frequency (a notch width of $hi \pm 5$ Hz). The ECoG signal was analyzed within the high-gamma frequency band (70–170 Hz). Significant ECoG activation values ($p < 0.001$), determined based on the Student t-distribution of the high-gamma band power, were used to test the null hypothesis of no correlation between band power samples and the cue of the performed task. Depending on the number of recorded high-gamma band power samples and the resultant R^2 value from the ECoG-FLM mapping procedure, the corresponding t-value indicated significant activation [34].

2.4.2 The "ESM-ML Guide" Approach. The Schematic Data Analysis Flow Is Presented in Fig. 4

The analysis was similar to that performed in our previous study with a Random Forest (RF) Classifier [30] and included the following steps:

(Step 1) All ECoG sites were divided into ESM positive [ESM(+)] and ESM negative [ESM(−)] (Table 2);

(Step 2) Features of ESM(+) and ESM(−) electrodes in the ECoG signal were determined by analyzing the signal in the frequency domain;

(Step 3) RF- and DL classifiers were trained to identify discriminative signal features in language-related ECoG activity;

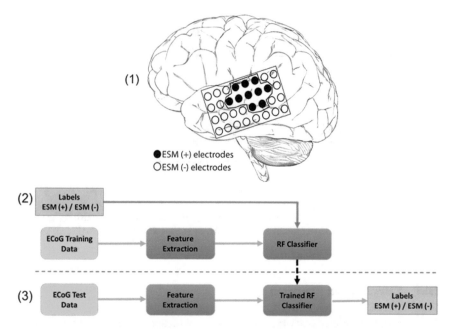

Fig. 4 Schematic representation of ESM-ML-guided analysis ("ESM-ML guide") of ECoG functional language mapping data: (1) ESM results are being used to separate the ECoG signal into two major categories: recorded from electrodes that were found to be ESM(+) and those that were found to be ESM(−); (2) Schematic representation of the workflow of machine learning analysis approach used in this study; (3) Future perspective of creating a "tool-box" with pre-loaded classifier trained on multiple language-related ECoG datasets, which can identify ESM(+) and ESM(−) features in prospectively recorded language-related ECoG activity (from Refs. [27–29])

Table 2 Description of ESM contacts identified in Step 1: ESM positive for language [ESM(+)] and ESM negative for language [ESM(−)]

Subject #	Number of ESM(+) contacts	Number of ESM(−) contacts	Total number of contacts tested with ESM
1	22	32	54
2	5	27	32
3	16	111	127
4	19	11	30
5	10	38	48
6	5	43	48

(Step 4) Accuracy of ESM-guided classification results was determined and compared with the accuracy of the conventional ECoG-FLM analysis approach;

(Step 5) Analysis was open to the whole spectrum of frequencies (0–600 Hz) and compared with high gamma analysis only (70–170 Hz) as well as with the analysis within alpha, beta, and gamma frequency bands combined together (8–170 Hz).

To test the stability of a machine learning classifier, a method of validation is often needed. A simple way of doing this is to train the classifier on a part of the data and then test it on the remaining data. This evaluation strategy is known as *hold-out* method. Based on this overall evaluation strategy, when there is limited data, as in our case, a more popular and bold approach is to use *k-fold cross-validation* method. In this validation approach, the data is split into k parts/folds (mostly equal samples in each fold), and the ML classifier is trained using $k - 1$ folds of the data and testing on the remaining fold. This is repeated k times until all folds have been tested exactly once and the results are averaged across the folds. For an example, in 10-fold cross validation approach, 9 folds of the data are used for training the ML classifier, while the remaining fold is used for testing. This operation continues for each fold of the data being separated for the test, and the remaining folds are being used in training. Since data folds are distinct, test data is never being used in training at the same time.

Again, similar to our previous classification approaches with RF [30], a 10-fold cross-validation was used here to test the accuracy of the newly established classifier in identifying channel response as ESM(+)/ESM(−) [30]. 10-fold cross validation results for the proposed DL approach are presented in the following section.

3   Results

The performances of the different approaches and classifiers were estimated using the standard metrics of accuracy, sensitivity and specificity. These metrics represented the percentage of correctly classified channels (i.e., accuracy), the percentage of positive channels identified correctly as positive (i.e., sensitivity) and the percentage of the number of negative channels identified correctly (i.e., specificity), respectively. The latter two metrics helped identify the strength of the classifier when there was imbalance in the data in terms of more positive channels than negative channels and vice-versa. These metrics were computed based on the channels.

The conventional ECoG-based approach yielded 58% accuracy, 22% sensitivity, and 78% specificity. The low sensitivity of the ECoG-based approach observed in this study is consistent with previously summarized results [17, 35]. It indicates the strong need for developing alternative strategies to solve this problem, which supports the rationale behind our current study. The proposed "ESM-ML guide" demonstrated significantly improved metrics compared to conventional ECoG-based approach. The "ESM-RF guide" (the use of RF classifier) showed 76.2% accuracy, 73.6% sensitivity, and 78.78% specificity, leading to a dramatic increase in the sensitivity of ECoG-FLM (73.6 vs. 22%). This prediction was further improved by the "ESM-DL guide" (using DL classifier), which yielded 83% accuracy, 84% sensitivity and 83% specificity (Fig. 5). With these rates, we achieved the state of the art values for sensitivity, specificity, and accuracy in identifying ESM(+) versus ESM(−) channel responses within the ECoG-FLM data.

We also compared the results obtained from analysis in different frequency bands against the results obtained analyzing the full spectrum of frequencies (Fig. 6). We

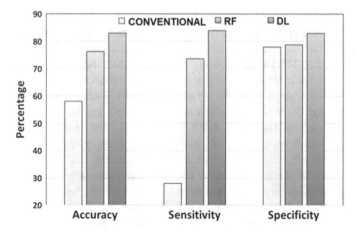

Fig. 5 Accuracy, sensitivity and specificity values when utilizing ESM-guided random forest (RF) and deep learning (DL) approaches (*"ESM-ML guides"*) compared with conventional ECoG analysis approach by using the whole spectrum of frequencies (from Refs. [27, 28])

Fig. 6 Classification scores for ECoG signal classification on language-related tasks. ABG—classification using α, β, γ bands combined (80–170 Hz), FS—classification using full signal spectrum (0–600 Hz) and G—classification using only γ band (70–170 Hz) (from Ref. [29])

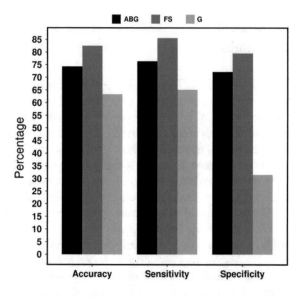

found that the lower frequency bands, specifically, alpha and beta, were not the main contributors towards high classification accuracy, whereas analysis within the high-gamma frequency band achieved relatively high classification accuracy. However, only the use of the full signal spectra (0–600 Hz) yielded the highest classification accuracy, sensitivity and specificity when compared to the "sub-band" approaches.

4 Discussion and Future Perspectives

ECoG-FLM is a functional mapping tool that can be of essential value in preventing post-surgical language deficits in epilepsy patients [14]. Current ECoG-FLM accuracy is not high enough compared with the gold standard of functional mapping—ESM. Despite relatively high specificity, ECoG-FLM lacks sensitivity [17, 35]. There are two possible ways for further ECoG-FLM development. The first route entails the use of ECoG-FLM with its current accuracy to complement ESM. Another route involves improving current ECoG-FLM accuracy, which could help ECoG-FLM evolve into an independent functional mapping modality.

4.1 ECoG-FLM as a Complimentary Tool for Functional Mapping

ECoG-FLM has been used as a complementary functional mapping tool to optimize the ESM in terms of time, effort effectiveness and safety (e.g., decreased probability of stimulation-induced seizures) [36–38]. Indeed, it has been shown that ECoG-based functional mapping helps identify potential ESM stimulation sites. This may result in a significant reduction ($39.8\%, p < 0.05$) of stimulated electrode pairs, hence reducing ESM time and improving ESM protocol [36, 38, 39].

ECoG-based functional mapping can also provide additional information that is not available through ESM. A number of ECoG-FLM-positive electrodes are not sampled by ESM due to various reasons [17, 37, 40]. For an example, 8 of the 12 patients studied had 17% ESM non-stimulated positive ECoG-FLM findings in addition to electrodes identified as positive for language via ESM [40] (Fig. 7). Importantly, the cortical sites underneath a portion of non-stimulated ECoG-FLM positive electrodes in some of these patients were surgically removed. Overall, several areas under the electrodes not sampled by ESM were found to be positive for language using ECoG-FLM. This shows additive value of ECoG-FLM to detect areas of functional significance. The findings of this study [40] indicated that surgical resection may be potentially compromising/sacrificing areas of the eloquent language cortex that are not always captured by ESM due to various reasons, including time constraints and/or ESM-provoked seizures.

The ESM value in predicting post-surgical morbidity has been a matter of recent discussions. A number of studies have demonstrated the failure of ESM to identify critical language sites [41–43]. At the same time, the sites that were falsely identified by ESM as negative for language were found as language positive by either ECoG-FLM, functional magnetic resonance imaging (fMRI) or both [44–48]. False negative results from ESM create a risk of surgical resection of areas essential for language, leading to potential post-surgical language deficits. This brings the added value of ECoG-FLM for the prediction of functional deficits for ESM negative sites to the level of major importance. Consequently, the use of ESM as a gold standard should be considered with caution, including its "ground truth" value for the comparison with ECoG-FML. Indeed, such comparison can produce low performance metrics for ECoG-FML (in terms of both sensitivity and specificity), obscuring the true value of the method.

Whereas ESM may better predict post-surgical deficits in some aspects of language function, the ECoG-FLM can be a better predictor of pre-surgical deficits in other functional language domains. A study of five patients with pharmacoresistant epilepsy [49] evaluated their ECoG-FLM and ESM mapping results, their neuropsychological post-surgical language outcome data, and established their surgical resection margins. A prediction was made on whether ECoG-FLM results were associated with surgical language outcome. The prediction was based on whether the brain tissue, located underneath the ECoG-FLM positive [ECoG-FLM(+)] electrodes, was removed. The same prediction was made based on ESM results. Afterwards, the predicted outcome (deficit/no deficit) was compared with real post-surgical language outcomes that became available after the post-operative neuropsychological evaluations (3–12 months after surgery). The results of this preliminary study [49] suggested that ECoG-FLM may predict an overall post-surgical language outcome to the similar extent as ESM. Importantly, ECoG-FLM may be a better predictor of post-surgical *receptive language outcome* than ESM. Conversely, ESM may be a better predictor of post-surgical *expressive language outcome*.

The future of using ECoG-FLM as a complementary functional mapping technique involves its potential to complement ESM (perhaps in one device, where ECoG-FLM "guides" ESM [50]) and, possibly, with other functional mapping modalities. The additive value of ECoG-FLM should be further explored [37]. Specifically, the implications of resecting the sites underlying ECoG-FLM(+) electrodes, which were

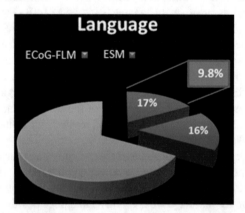

Fig. 7 Demonstration of additive ECoG-FLM value. 16% of total subdural electrodes placed in 8 of the 12 patients were *ESM positive for language*. 17% of electrodes that were not stimulated by ESM were found *positive for language function by ECoG-FLM*. Approximately 9.8% of these latter electrode locations were subsequently resected, potentially leading to post-surgical language deficits (from Ref. [40])

not evaluated by ESM, should be further explored in relation to the results of post-surgical neuropsychological evaluations. Following this correlative analysis, clinical significance of the ECoG-FLM's additive potential will be determined. Additive values of other imaging modalities, such as fMRI and magnetoencephalography (MEG) should be established as well. Furthermore, more comprehensive predictive models of post-surgical language outcomes, based on ECoG-FLM and ESM results, should be developed [49]. They may focus on the magnitude of the significance of the ECoG-FLM(+) or ESM(+) electrodes in the resection zone, as well as on their number. Moreover, not only a physiological, but also a clinical threshold must be taken into consideration, for example, when evaluating post-surgical language outcome values. Finally, relative risks of type 1 and 2 errors must be considered.

4.2 ECoG-FLM as an Independent Functional Mapping Modality

Although ECoG-FLM appears most promising as a complementary functional mapping modality in the near future, there is still a possibility for it to become an independent imaging modality if it can at least reproduce results obtained by ESM. We believe that ECoG-FLM "reproduction" accuracy of ESM results can be significantly increased by: (1) Using ESM as a guiding tool; (2) Employing analysis methods derived from the artificial intelligence domain (e.g., Deep Learning); and (3) Changing the conventional ECoG-FLM analysis bandwidth (usually restricted within the high-gamma spectrum of frequencies).

In our current study, we aimed to demonstrate the feasibility of utilizing an ESM-guided approach for obtaining ECoG-FLM results in six patients with pharmacoresistant epilepsy. Similar to our previous study [30], we applied the ML feature extraction methodology to the already available ECoG-FLM data and estimated functional mapping accuracy within the same dataset. In order to increase our previously achieved classification accuracy with a Random Forest Classifier [30], we employed a Deep Learning Classifier. Our preliminary results from six patients revealed that the "ESM-DL guide" exceeded both the conventional ECoG-FLM data analysis approach and "ESM-RF guide" in terms of accuracy, sensitivity and specificity.

The low concordance between ECoG-FLM and ESM might reflect the complexity of language function and its cortical representation [51]. Therefore, it is possible to speculate that some features of the ECoG signal, reflecting complex nature of language processing, are omitted from consideration when restricting the ECoG-FLM analysis to the gamma frequency band only (a conventional approach that is currently utilized). There are several studies to support this view. For example, Strauss, Kotz [52] have indicated that power increases in both theta- and alpha- activity accompany speech recognition. The pivotal role of theta activity in language comprehension has been demonstrated [53]. Contributions of the lower frequencies to increased sensitivity and of the higher frequencies to increased specificity of ECoG-FLM results have also been noted [10]. Importantly, as previously suggested [17], the bias towards higher specificity of ECoG-FLM can be explained by limiting the ECoG analysis to high frequency range only (e.g., high gamma). According to the current findings in language research, in order to comprehensively portray the complexities of language representation in the brain, joint analysis of the whole range of frequency bands and their detailed analysis is essential. Therefore, in addition to using ESM to guide ECoG-FLM, we have opened our analysis to a wide range of frequencies (0–600 Hz) [30]. As expected, the analysis results in the gamma frequency band were superior to those in lower bands only in terms of ECoG-FLM accuracy (Fig. 6). Opening analysis to the whole spectrum of frequencies has exceeded the results gained from prior analysis approaches (Fig. 6). It contributed to the improved classification accuracy and confirmed the results of previous studies, pointing towards the complex nature of language processing that needs to be considered during the analysis of neurophysiological data. Importantly, unlike previous review reports [17, 35], there was no inclination for the significant bias towards increased specificity in ECoG-FLM results (the results for DL classifier demonstrated 83% accuracy, 84% sensitivity and 83% specificity).

Although the analysis in lower frequency range should not be underestimated, the importance of analyzing complex neurophysiological data in the range of High Frequency Oscillations (HFOs) (70 Hz and higher) is of particular value [54]. Our results have demonstrated that the analysis of fast ripples (250–500 Hz) (which is much higher than high-gamma frequency band used for ECoG-FLM) can allow discrimination between the ECoG signals recorded from ESM(+) and ESM(−) channels [30] (Fig. 8).

Importantly, some of these differences were still detectable in a control task condition (Fig. 9), possibly indicating the intrinsic nature of the observed differences in ECoG signal recorded from cortical tissue underlying ESM(+) and ESM(−) electrodes. Further understanding of these processes may lead to identification of functional language networks from recorded ESM(+) and ESM(−) ECoG activity, potentially giving rise to a new passive functional mapping methodology (for example, see [55]).

4.3   Summary and Future Developments

In conclusion, we have demonstrated for the first time that it is possible to improve current ECoG-FLM accuracy by using the ESM-navigated approach ("ESM-FLM guide") supported by advanced machine learning algorithms, such as Deep Learning. This shows the feasibility of using ECoG-FLM as a stand-alone functional mapping modality. Future studies will be aiming at increasing study sample size and improving ML classification parameters. Importantly, a validation with the prospective clinical data is warranted. The long-term goal of this project is to create a toolbox with a "ready to use ESM-ML guide" algorithm trained to provide high accuracy ECoG-FLM results by classifying between ESM positive and ESM negative

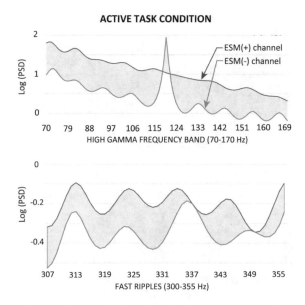

Fig. 8 The Power Spectral Density (PSD) of the ECoG signal recorded during the *active task condition* from the channels that were found ESM(+) *(violet color)* and ESM(−) *(dark green color)* for language. **Top**: PSD was estimated in high gamma frequency range (70–170 Hz); **Bottom**: PSD was estimated in fast ripple frequency range (300–355 Hz). The difference between the PSD calculated for ESM(+) and ESM(−) ECoG channels is shaded in *gray* (adopted from [30])

results in prospective sets of language-related ECoG data. We believe that implementing such easy-to-use and reliable technology will increase current pre-surgical and intra-operative functional mapping accuracy, thus preventing post-surgical language morbidity and improving patient outcomes. Available software packages for neurophysiological data analysis, such as "MEG Processor" (https://sites.google.com/site/braincloudx/), developed by J. Xiang, already proved useful in analyzing neurophysiological data within an unprecedented frequency range of 1–2884 Hz at the source levels using newly developed analysis approaches [56]. Such software can potentially serve as the main building blocks for "ESM-ML guide" software package development. Such approaches are especially important due to the value brought by the analysis of HFOs (high gamma activity, ripples, and fast ripples).

Further steps will entail the transition of the "ESM-ML guide" from off-line analysis of ECoG-FLM signal into a real-time tool to be used intra-operatively. Our current study is the first step towards the development of this clinically relevant software. Amending such software with ecological paradigms, allowing automated functional brain mapping during simultaneous video and ECoG recordings in patients undergoing an invasive evaluation for epilepsy surgery, is warranted (for an example of such development for functional motor mapping, see [57]). An important contribution of our novel approach is that it can be applied to non-invasive functional mapping

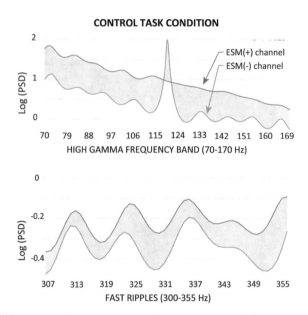

Fig. 9 The Power Spectral Density (PSD) of the ECoG signal recorded during the *control condition* from the channels that were found ESM(+) *(blue color)* and ESM(−) *(light green color)* for language. **Top**: PSD was estimated in high gamma frequency range (70–170 Hz); **Bottom**: PSD was estimated in fast ripple frequency range (300–355 Hz). The difference between the PSD calculated for ESM(+) and ESM(−) ECoG channels is shaded in *gray* (adopted from [30])

modalities, such as MEG or fMRI, to improve their accuracy and reduce the number of invasive recordings. Seizure detection in ongoing neurophysiological activity during the ECoG-FLM is also recommended.

Acknowledgements The authors acknowledge The Central Florida Health Research (CFHR) grant (PIs: Drs. M. Korostenskaja and U. Bagci) for supporting this study. Dr. Korostenskaja also would like to express her gratitude to Drs. G. Schalk and W. Wang for their feed-back/criticism/contribution to the development of this project's idea for submission to NIH.

References

1. G. Schalk et al., Brain-computer interfaces (BCIs): detection instead of classification. J. Neurosci. Methods **167**(1), 51–62 (2008)
2. G. Schalk et al., Real-time detection of event-related brain activity. Neuroimage **43**(2), 245–249 (2008)
3. R. Arya et al., Presurgical language localization with visual naming associated ECoG high—gamma modulation in pediatric drug-resistant epilepsy. Epilepsia **58**(4), 663–673 (2017)
4. C. Kapeller et al., CortiQ-based real-time functional mapping for epilepsy surgery. J. Clin. Neurophysiol. **32**(3), e12–22 (2015)

5. M. Korostenskaja et al., Real-time functional mapping with electrocorticography in pediatric epilepsy: comparison with fMRI and ESM findings. Clin. EEG Neurosci. **45**(3), 205–211 (2014)
6. T. Kambara et al., Presurgical language mapping using event-related high-gamma activity: The Detroit procedure. Clin. Neurophysiol. **129**(1), 145–154 (2018)
7. Y. Tamura et al., Passive language mapping combining real-time oscillation analysis with cortico-cortical evoked potentials for awake craniotomy. J. Neurosurg. **125**(6), 1580–1588 (2016)
8. Y. Cho-Hisamoto et al., Cooing- and babbling-related gamma-oscillations during infancy: intracranial recording. Epilepsy Behav. **23**(4), 494–496 (2012)
9. R. Arya et al., Electrocorticographic language mapping in children by high-gamma synchronization during spontaneous conversation: comparison with conventional electrical cortical stimulation. Epilepsy Res. **110**, 78–87 (2015)
10. P.R. Bauer et al., Mismatch between electrocortical stimulation and electrocorticography frequency mapping of language. Brain Stimul. **6**(4), 524–531 (2013)
11. K. Kamada et al., Disconnection of the pathological connectome for multifocal epilepsy surgery. J. Neurosurg. 1–13 (2017)
12. H. Ogawa et al., Rapid and minimum invasive functional brain mapping by realtime visualization of high gamma activity during awake craniotomy. World Neurosurg. (2014)
13. N.E. Crone, A. Sinai, A. Korzeniewska, High-frequency gamma oscillations and human brain mapping with electrocorticography. Prog. Brain Res. **159**, 275–295 (2006)
14. M. Korostenskaja et al., Real-time functional mapping: potential tool for improving language outcome in pediatric epilepsy surgery. J. Neurosurg. Pediatr. **14**(3), 287–295 (2014)
15. P. Brunner et al., A practical procedure for real-time functional mapping of eloquent cortex using electrocorticographic signals in humans. Epilepsy Behav. **15**(3), 278–286 (2009)
16. C.D. Wray et al., Multimodality localization of the sensorimotor cortex in pediatric patients undergoing epilepsy surgery. J. Neurosurg. Pediatr. **10**(1), 1–6 (2012)
17. M. Korostenskaja et al., Electrocorticography-based real-time functional mapping for pediatric epilepsy surgery. J. Pediatr. Epilepsy **04**(04), 184–206 (2015)
18. X. Zhang et al., Surgical treatment for epilepsy involving language cortices: a combined process of electrical cortical stimulation mapping and intra-operative continuous language assessment. Seizure **22**(9), 780–786 (2013)
19. G. Ojemann et al., Cortical language localization in left, dominant hemisphere. An electrical stimulation mapping investigation in 117 patients. J. Neurosurg. **71**(3), 316–26 (1989)
20. S. Visa, A. Ralescu, Data-driven fuzzy sets for classification. Int. J. Adv. Intell. Paradigms **1**(1), 3–30 (2008)
21. T.M. Rutkowski et al., Multichannel spectral pattern separation—an EEG processing application. in *IEEE International Conference on Acoustics, Speech and Signal Processing, 2009. ICASSP 2009* (2009)
22. W. Wang et al., An electrocorticographic brain interface in an individual with tetraplegia. PLoS ONE **8**(2), e55344 (2013)
23. C. Kapeller et al., Single trial detection of hand poses in human ECoG using CSP based feature extraction. Conf. Proc. IEEE Eng. Med. Biol. Soc. **2014**, 4599–4602 (2014)
24. M. Korostenskaja et al., Improving ECoG-based P300 speller accuracy, in *Proceedings of the 6th International Brain-Computer Interface Conference 2014*, vol. 088 (2014), pp. 1–4
25. G. Schalk, E.C. Leuthardt, Brain-computer interfaces using electrocorticographic signals. IEEE Rev. Biomed. Eng. **4**, 140–154 (2011)
26. A. Ralescu, K.H. Lee, M. Korostenskaja, Machine learning techniques provide with a new insight into pre-attentive information processing changes in pediatric intractable epilepsy, in *Society for Psychophysiological Research 51st Annual Meeting 2011* (Boston, MA, USA, 2011), p. 92
27. M. Korostenskaja et al., Gold standard for epilepsy/tumor surgery coupled with deep learning offers independence to a promising functional mapping modality, in *Submission for Annual International BCI2017 Award (Nominee for 2017:* http://www.bci-award.com/2017) (2017)

28. M. Korostenskaja et al., Gold standard for epilepsy/tumor surgery coupled with deep learning offers independence to a promising functional mapping modality, in *Poster presentation at the Annual International BCI2017 Award ceremony during the 7th Graz Brain Computer Interface Conference 2017, Graz, Austria*, 18–22 September 2017

29. M. Korostenskaja et al., ESM-guided approach supported by machine learning improves accuracy of ECoG-based functional language mapping, in *American Epilepsy Society Annual Meeting 2017* (Washington, DC, USA, 2017), p. Abst. 1.110

30. H. RaviPrakash et al., Automatic response assessment in regions of language cortex in epilepsy patients using ECoG-based functional mapping and machine learning, in *2017 IEEE International Conference on Systems, Man, and Cybernetics (SMC)* (2017), pp. 519–524

31. M. Korostenskaja et al., Characterization of cortical motor function and imagery-related cortical activity: potential application for prehabilitation, in *2017 IEEE International Conference on Systems, Man, and Cybernetics (SMC)* (2017), pp. 3014–3019

32. G. Schalk et al., BCI2000: a general-purpose brain-computer interface (BCI) system. IEEE Trans. Biomed. Eng. **51**(6), 1034–1043 (2004)

33. C. Kapeller et al., cortiQ—based real-time functional mapping for epilepsy surgery. J. Clin. Neurophysiol. **32**(3), e12–22 (2015)

34. N.A. Rahman, A course in theoretical statistics: for sixth forms, technical colleges, colleges of education, universities (Charles Griffin & Company Limited, 1968)

35. R. Arya, P.S. Horn, N.E. Crone, ECoG high-gamma modulation versus electrical stimulation for presurgical language mapping. Epilepsy Behav. **79**, 26–33 (2018)

36. R. Prueckl et al., Passive functional mapping guides electrical cortical stimulation for efficient determination of eloquent cortex in epilepsy patients. in *IEEE Biomedical Conference 2017 Proceedings* (2017)

37. M. Elsayed et al., Additive Potential of Real-Time Functional Mapping (RTFM) to Electrical Stimulation Mapping (ESM) results for epilepsy surgery candidates, in *American Epilepsy Society Meeting 2014* (Seattle, Washington, USA, 2014), p. 3.276

38. R. Prueckl et al., O202 Combining the strengths of passive functional mapping and electrical cortical stimulation. Clin. Neurophysiol. **128**(9), e243 (2017)

39. C. Kapeller et al., Electrocorticography guides electrical cortical stimulation to identify the eloquent cortex, in *American Epilepsy Society Annual Meeting 2017* (Washington, DC, USA, 2017), p. Abst. 1.123

40. M. Elsayed et al., Additive Potential of Real-Time Functional Mapping (RTFM) to Electrical Stimulation Mapping (ESM) Results for Epilepsy Surgery Candidates, in *Poster for American Epilepsy Society Meeting 2014* (Seattle, Washington, USA, 2014)

41. K.G. Davies, G.L. Risse, J.R. Gates, Naming ability after tailored left temporal resection with extraoperative language mapping: increased risk of decline with later epilepsy onset age. Epilepsy Behav. **7**(2), 273–278 (2005)

42. B.P. Hermann et al., Visual confrontation naming following left anterior temporal lobectomy: a comparison of surgical approaches. Neuropsychology **13**(1), 3–9 (1999)

43. M.J. Hamberger et al., Brain stimulation reveals critical auditory naming cortex. Brain **128**(Pt 11), 2742–2749 (2005)

44. M. Genetti et al., Comparison of high gamma electrocorticography and fMRI with electrocortical stimulation for localization of somatosensory and language cortex. Clin. Neurophysiol. **126**(1), 121–130 (2015)

45. K. Kojima et al., Gamma activity modulated by picture and auditory naming tasks: intracranial recording in patients with focal epilepsy. Clin. Neurophysiol. **124**(9), 1737–1744 (2013)

46. K.J. Miller et al., Rapid online language mapping with electrocorticography. J. Neurosurg. Pediatr. **7**(5), 482–490 (2011)

47. A. Sinai et al., Electrocorticographic high gamma activity versus electrical cortical stimulation mapping of naming. Brain **128**(Pt 7), 1556–1570 (2005)

48. M.C. Cervenka et al., Language mapping in multilingual patients: electrocorticography and cortical stimulation during naming. Front. Hum. Neurosci. **5**, 13 (2011)

49. M. Korostenskaja et al., Predicting post-surgical language outcome with ECoG-based real-time functional mapping (RTFM) in patients with pharmaco-resistant epilepsy, in *American Epilepsy Society 68th Annual Meeting 2014* (Seattle, Washington, USA, 2014), p. 2.255
50. R. Prueckl et al., Passive functional mapping guides electrical cortical stimulation for efficient determination of eloquent cortex in epilepsy patients, in *2017 39th Annual International Conference of the IEEE Engineering in Medicine and Biology Society (EMBC)* (2017)
51. G. Hickok, D. Poeppel, The cortical organization of speech processing. Nat. Rev. Neurosci. **8**(5), 393–402 (2007)
52. A. Strauss et al., Alpha and theta brain oscillations index dissociable processes in spoken word recognition. Neuroimage **97**, 387–395 (2014)
53. E. Halgren et al., Laminar profile of spontaneous and evoked theta: rhythmic modulation of cortical processing during word integration. *Neuropsychologia* (2015)
54. J. Xiang et al., High frequency oscillations in pediatric epilepsy: methodology and clinical application. J. Pediatr. Epilepsy **04**(04), 156–164 (2015)
55. A.L. Ko et al., Identifying functional networks using endogenous connectivity in gamma band electrocorticography. Brain Connect **3**(5), 491–502 (2013)
56. J. Xiang et al., Multi-frequency localization of aberrant brain activity in autism spectrum disorder. Brain Dev. **38**(1), 82–90 (2016)
57. P. Gabriel et al., Neural correlates to automatic behavior estimations from RGB-D video in epilepsy unit. Conf. Proc. IEEE Eng. Med. Biol. Soc. **2016**, 3402–3405 (2016)

Online Adaptive Synchronous BCI System with Attention Variations

Susan Aliakbaryhosseinabadi, Ernest Nlandu Kamavuako, Ning Jiang, Dario Farina and Natalie Mrachacz-Kersting

Abstract In real-life scenarios, outside of the laboratory setting, the performance of brain-computer interface (BCI) systems is influenced by the user's mental state such as attentional diversion. Here, we propose a novel online BCI system able to adapt with variations in the users' attention during real-time movement execution. Electroencephalography (EEG) signals were recorded from twelve channels in twelve healthy participants and two stroke patients while performing 50 trials of ankle dorsiflexion simultaneously with an auditory oddball task. For each participant, the selected channels, classifiers and features from the offline mode were used in the online mode to predict the attention status. For both healthy controls and subacute stroke patients, feedback to the user on attentional status reduced the amount of attentional diversion created by the oddball task. The findings presented here demonstrate that the users' attention can be monitored in a fully online BCI system, and further, that real-time neurofeedback on the attentional state of the user can be implemented to focus the attention of the user back onto the main task of the BCI for neuromodulation. Monitoring the users' attention status will have a major impact in the BCI for neurorehabilitation area in the future.

Keywords Brain-computer interface · Attentional diversion · Real-time feedback · EEG feature extraction

S. Aliakbaryhosseinabadi · N. Mrachacz-Kersting (✉)
Center for Sensory-Motor Interaction, Department of Health Science and Technology, Aalborg University, 9220 Aalborg, Denmark
e-mail: nm@hst.aau.dk

E. N. Kamavuako
Department of Informatics, King's College London, London, UK

N. Jiang
Department of System Design Engineering, Faculty of Engineering, University of Waterloo, Waterloo, Canada

D. Farina
Department of Bioengineering, Imperial College London, SW7 2AZ London, UK

© The Author(s), under exclusive licence to Springer Nature Switzerland AG 2019   31
C. Guger et al. (eds.), *Brain-Computer Interface Research*,
SpringerBriefs in Electrical and Computer Engineering,
https://doi.org/10.1007/978-3-030-05668-1_3

1   Introduction

Attention is the ability to process resources related to a particular sensory stimulus, memory, thought or any other mental task [1, 2]. Attention can be influenced by either external sensory stimuli such as sounds, images or smells (exogenous attention) or internal events such as memories or thoughts (endogenous attention). Task behaviors including reaction time or accuracy of movement detection are affected by the amount of attention to the task, and the level of attention is a key factor underlying variations in task execution [3].

Typically, four main types of attention have been proposed—sustained attention, alternative attention, selective attention and divided attention (Fig. 1). Sustained attention is the ability to focus on a particular task for a prolonged duration of time, such as reading a book. Training of sustained attention improves reaction time to the target task [4]. Alternative attention represents the flexibility of the brain to switch between two tasks with different cognitive demands such as reading a recipe and cooking a meal. Patients that suffer from childhood chronic fatigue syndrome show an increase in mental fatigue when asked to switch attention between two tasks [5, 6]. The ability to select a desired stimulus among various stimuli in the environment is referred to as selective attention and has been investigated via the cocktail-party effect - the ability to listen to a target speech in a noisy environment. The cortical activation corresponding to the target speech selected by selective attention is increased as compared to the non-selective attention conditions [7]. Divided attention refers to the ability to do two or more tasks concurrently such as dual-tasking. Attention is divided among two tasks, which has an effect on task accuracy and reaction time [8, 9].

Attentional distraction from main task execution leads to costs for the attended task, which can be expressed in the form of increments in the reaction time or deterioration in movement preparation [10]. For example, if the doctor's attention is distracted during surgery due to outside distractions, it will influence the doctor's performance such as the correct selection of instruments [11].

Out of these main types of attention, studies have investigated the effect of divided attention by asking participants to perform dual tasks [12, 13]. Dual tasking is defined

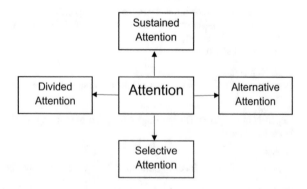

Fig. 1 Attention may be divided into four types: sustained, alternative, selective and divided attention

as the execution of two tasks simultaneously. Since attention is divided between two tasks in dual task conditions, the level of attention to the main task is reduced as compared to single task execution, and thus the other task can be considered a distractor for the main task execution.

2 Brain-Computer Interfaces and Attention

Performance of BCI systems is influenced by factors such as the users' attentional variations, fatigue and task learning. Previous studies have shown that an increment in the subject's level of fatigue, as well as a reduction in the attention level, both lead to lower task performance [14, 15]. The effect of attention alteration using various types of stimuli (visual, auditory and audiovisual) has been investigated by analyzing EEG signal characteristics obtained from different signal modalities such as the event-related potential (ERP) or steady-state evoked potentials (SSVEP) [16, 17]. These demonstrate that attention can be monitored by analyzing EEG signals obtained from different brain regions responsible for controlling attention to the task. For instance, a visual stimulus in the form of a bar used to control eye movement alters the P300 amplitude in fronto-central and occipital regions [16, 17]. Typically, ERP components such as the P300 latency and amplitude [18] are extracted to quantify the level of the attention. The P300 is a positive peak that occurs 300–400 ms after the onset of the stimulus that modulates the attention level. If attention to the stimulus is high, the P300 amplitude is increased and the P300 latency is decreased [19, 20].

BCIs have also been devised to monitor each user's attention to the target motor task to recognize attentional distraction [8, 9, 20] with the end goal of guiding the attention back to the task. In these studies, the distractor was in the form of a cognitive task cued by an auditory oddball paradigm where participants were asked to count the number of pre-defined target sequences of auditory tones. The main task was ankle dorsiflexion timed to a visual cue. Temporal features of the movement-related cortical potential (MRCP) were extracted, and it was demonstrated that these may be used to quantify the attention level of the users. Thus, pre-movement slopes in the time window of [1 0] s prior to movement execution and the value of the peak negativity of the MRCP decreased if the attention to the main task was reduced due to dual-tasking. The final goal of these series of studies is to provide feedback to the users to compensate for attention shifts from the main BCI tasks, since the users' mental states like fatigue and attention influence BCI performance [22]. In the past, we have mainly applied our MRCP based BCI to induce plasticity in patient populations such as those suffering from stroke. According to the Hebbian principle, synapses are strengthened and routes are formed among neurons when they are activated in a correlated fashion [23]. The neurofeedback in the form of an electrically induced sensory feedback was effective in inducing plasticity only if it was timed to arrive at the motor cortex during the peak negative phase of the movement-related cortical potential (MRCP) generated by the task imagination or execution [24]. Here, attention to the task is vital, as it is well known that attention plays a key role in plasticity induction as low attention to the task causes no plasticity induction [25], regardless of the timing between the movement intention and external trigger [25, 26]. Plasticity induction is required for functional recovery through neurorehabilitation following stroke [27].

Among different types of attention (sustained, alternate, selective and divided attention), alternate and divided attention are more critical for BCIs used in the clinical setting. For example, distractors in the form of noise from the surrounding environment cause the user to divide attention during the execution of a particular exercise, such as a fellow patient coming into the training room wishing to say hello.

3 Dual Tasking

Dual tasking is the ability to execute two tasks simultaneously, and leads to a division of attention between the two tasks [12, 13]. In dual-tasking, one of the tasks plays the role of interference (secondary task) for the other task (main task) and degrades the performance of the main task in comparison to single-task conditions. The amount of attention allocated to each task among various types of tasks, called prioritization, is specified by a tradeoff between task costs, since improvement in one task corresponds to deterioration in the other task. Reaction time and error rate are two examples of dual-task costs [28]. One of the possible reasons of these costs is the brain's limited ability for information processing. Based on different task information, various parts of the brain are activated, and thus competition between task resources leads to task costs [29, 30]. These are enhanced in the case of incompatibility between either task responses or task stimuli [31]. In case of compatibility, the same cortical parts of the brain are activated to respond to the tasks. If two tasks with different stimuli need similar responses such as pressing a button, it can reduce the costs. In the same way, compatible task stimuli (same color or shape) with different responses (pressing two different buttons) influence task costs.

Dual-tasking costs have been quantified by changes in response time and error rate of task execution [32]. For example, the number of errors of a cognitive task (counting of auditory tones) increased when performed simultaneously with a motor task [8]. These costs may be explained by the competition between task resources as each task activates corresponding cortical sites without significant overlap with other areas [32]. Based on this, each task is analyzed by switching between two task resources for information analysis [33]. Others have shown that, while a motor task execution (walking) is performed concurrently with a cognitive task (counting), the dual-task costs will increase because the motor task preparation deteriorates due to attention division between two concurrent tasks [34, 35]. In these studies, participants were asked to walk on a treadmill at either a self-selected or fast speed with and without doing a cognitive task. In the dual-task condition, breathing rate was increased significantly while gait speed was decreased compared to the single-task level. These findings demonstrate that cognitive load during walking (dual-tasking) has both metabolic and performance costs.

Dual-task execution is more critical for elderly populations when they perform tasks such as walking while talking at the same time, as this has been shown to increase the possibility of falling [36]. Thus, in addition to task resource competition, aging increases the reaction time and error rates for dual-tasks as compared to single-tasks [37, 38]. If the skills of the elderly in task execution are increased by practicing (training), the task performance will be increased [39, 40].

Stroke patients have more difficulties in dual-task conditions due to their age and also their disabilities. In addition, fatigue is more critical for stroke patients, especially in the acute and subacute phase [41]. Overall, attentional state is changed in dual-task conditions based on the aforementioned reasons of task prioritization, task compatibility, aging and fatigue. A BCI designed for neurorehabilitation should be able to monitor attention in real time and detect attention shifts during task execution. In this way, it is possible to alert the therapist (and/or the patient) in case of attentional drift and compensate for the resulting costs.

4 Feedback During Attention Alteration

The current work focused on quantifying the effect of artificially generated attention shifts on the performance of the main motor task and various temporal and spectral EEG features. Up to now, previous studies have attempted to identify EEG signal characteristics by using various types of attention distractors (visual, auditory, audio-visual) during main task execution. These studies used EEG signal modalities such as ERPs and functional magnetic resonance imaging (fMRI) from different parts of the brain as a control signal to monitor attentional states to the task [16, 17, 42]. Our previous studies attempted to investigate the effect of reducing attention to the main task (motor task execution) by using a secondary task (cognitive task of counting) in the offline mode. In the current study, the aim was to recognize distraction from the target task in the online mode. The goal of controlling attentional state is to use this information to provide feedback to the users to drive their attention back to the main BCI task.

The experimental design (Fig. 2) consisted of an offline calibration phase and an online phase. For the online phase, one group of subjects received feedback on their attentional state while no such information was provided to the other group. In all phases, subjects were asked to perform a main task of dorsiflexion of the dominant foot timed to a visual cue played on a screen approximately 1.5 m in front of the subject (single-task condition). The cue consisted of focus, preparation, execution, hold and rest phases. In the offline calibration phase, subjects completed a total of 50 such movements. Interspersed with these, random trials were provided in addition to the visual cue, consisting of an auditory oddball task (diverted attention trials - dual-task level). In these, participants had to focus on both tasks at the same time, i.e. to execute ankle dorsiflexion and also to count the number of desired sequences played in the oddball. EEG trials were obtained according to the movement onset extracted from electromyography (EMG) signals. Each trial was analyzed in the time and frequency

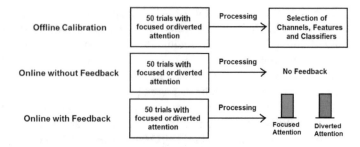

Fig. 2 Illustration of experimental procedure during training and online phases

domain to provide temporal and spectral features. Temporal features were extracted from the pre-movement phase of the MRCP signals, which indicate movement preparation and characteristics of each user's mental state. This followed from our previous studies where the MRCP was used to decode movement characteristics such as force and speed [43, 44] in addition to attention variations [21, 45, 46]. Pre-movement slopes and pre-movement variability in different time domains as well as amplitude and time of peak negativity were obtained from single MRCP trials. Spectral features obtained from five frequency band powers [Delta (0–3 Hz), Theta (4–7 Hz), Alpha (8–12 Hz), Beta (13–32 Hz) and Gamma (33–80 Hz)] served as additional features for attention classification. This feature group has previously been used in EEG-based BCIs to classify baseline EEG from a mental task and resulted in a classification accuracy of up to 97% [47]. Finally, a combination of temporal and spectral features (tempo-spectral features) were implemented as these have been shown to improve classification of the users' attention [9]. The classifier was trained on the data from the offline phase using both the focused and diverted attention trials and the best classifier, features and channels were selected and used in the subsequent online phase.

In the online mode, each trial was classified and feedback was provided based on the output of the classification, as illustrated in Fig. 2. The green bar represents the focused attention trials, while the red bar indicates the diverted attention trials. To be able to investigate the effect of feedback in controlling attention level, subjects were divided into two groups, where one group received feedback during the online phase while no feedback was provided for the other group. Participants in the online mode with feedback attempted to increase the number of focused attention trials by enhancing the number of green bars.

Figure 3a illustrates the grand average MRCP for one participant for the focused and diverted attention trials. The time of peak negativity as well as pre-movement slopes decreased in the diverted attention level. Movement preparation amplitude was decreased due to attention reduction to the main task. Across all subjects, the results revealed significant differences in classification accuracy among the experimental phases of offline calibration, online phase with feedback and online phase without feedback. There was a significant decrement in accuracy between calibration and with-feedback phases (72.3 ± 4.3 to $60.1 \pm 7.9\%$) and also between without-feedback and with-feedback phases (67.1 ± 6.1 to $60.1 \pm 7.9\%$).

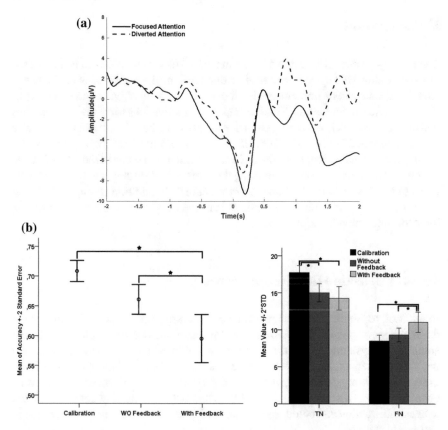

Fig. 3 The results of **a** grand average MRCP for two attention levels for one sample participant and **b** accuracy (left column) and TN and FN (right column) in different phases across all subjects

Significant differences were also observed for the rate of true negatives (TN—the number of focused attention trials classified correctly) between the experimental phases. There was a significant decrement from calibration to with-feedback phase (17.75 ± 2.3 to 14.3 ± 3.9) as well as a reduction from calibration to without feedback levels (17.75 ± 2.3 to 15 ± 3). False negative (FN—the number of diverted attention trials classified as focused attention values), revealed a significant difference between the experimental conditions. FN was significantly increased from the calibration to with-feedback phase (8.5 ± 2 to 11 ± 3.3) and between without feedback and with feedback conditions (9.3 ± 2.3 to 11 ± 3.3). Figure 3b illustrates the results obtained from the three different phases of the experiment.

5   Conclusion

The findings presented here demonstrate that a BCI user's attention can be monitored in a fully online BCI system, and further, that real-time neurofeedback on the user's attentional state can be implemented to focus the user's attention on the main task of the BCI for neuromodulation. This has important implications for patient populations that have difficulties in executing tasks due to attention alteration. It can also increase the performance of assistive BCI devices such as wheelchairs or robotic prosthesis by helping users remain focused on the main task. As attention is one of the vital effective factors on BCI performance, monitoring the users' attention status will have a major impact in the neurorehabilitation area in the future. In addition to implications for real-time BCI control, this work could bolster the development passive monitoring tools for much broader applications.

6   Jury Selection and Recent Work

In addition to implications for real-time BCI control, this work could bolster the development of passive monitoring tools for much broader applications. BCIs and similar EEG systems to monitor attention (and influence real-time systems accordingly) have been explored for decades, across user groups including pilots, drivers, astronauts, military personnel, gamers, and persons seeking rehabilitation for a variety of attentional disorders (e.g., [47–49]). The MRCP has usually not been utilized to assess distraction and could lead to more informative measurement of attention in real-world settings.

This year, one of us (NMK) was honored to serve as the chair of the BCI Award Jury. Consistent with earlier years, NMK did not participate in voting on this project to avoid a conflict of interest. We had a large jury this year, with six other members, and we were happy to be nominated based on their votes. Our team had several projects nominated for earlier BCI Awards, and we even won third place in 2014.

Among other recent work, one emerging direction involves ECoG activity from electrodes implanted in the brain during neurosurgery. ECoG activity can provide greater spatial resolution, frequency range, and artifact robustness than EEG. Capitalizing on this richer information has improved our understanding of how different brain regions contribute to attention [50–52]. This could in turn inspire improved methods for non-invasive methods such as the EEG.

References

1. M. Esghaei, M.R. Daliri, Decoding of visual attention from LFP signals of macaque MT. PLoS ONE **9**(6), 1–10 (2014)
2. M.S. Treder, B. Blankertz, (C)overt attention and visual speller design in an ERP-based brain-computer interface. Behav. Brain Funct. **6**(1), 28 (2010)
3. D. Purves, *Principles of Cognitive Neuroscience* (W. H. Freeman, New York, 2008)
4. A. Lutz, H.A. Slagter, N.B. Rawlings, A.D. Francis, L.L. Greischar, R.J. Davidson, Mental training enhances attentional stability: neural and behavioral evidence. J. Neurosci. **29**(42), 13418–13427 (2009)
5. J. Kawatani, K. Mizuno, S. Shiraishi, M. Takao, T. Joudoi, S. Fukuda et al., Cognitive dysfunction and mental fatigue in childhood chronic fatigue syndrome—a 6-month follow-up study. Brain Dev. **33**(10), 832–841 (2011)
6. K. Mizuno, M. Tanaka, S. Fukuda, K. Imai-Matsumura, Y. Watanabe, Relationship between cognitive functions and prevalence of fatigue in elementary and junior high school students. Brain Dev. **33**(6), 470–479 (2011)
7. Y. Gao, Q. Wang, Y. Ding, C. Wang, H. Li, X. Wu et al., Selective attention enhances beta-band cortical oscillation to speech under cocktail-party listening conditions. Front. Hum. Neurosci. **11**, 34 (2017)
8. S. Aliakbaryhosseinabadi, E.N. Kamavuako, N. Jiang, D. Farina, N. Mrachacz-Kersting, Influence of dual-tasking with different levels of attention diversion on characteristics of the movement-related cortical potential. Brain Res. **1674**, 10–19 (2017)
9. S. Aliakbaryhosseinabadi, E.N. Kamavuako, N. Jiang, D. Farina, N. Mrachacz-Kersting, Classification of EEG signals to identify variations in attention during motor task execution. J. Neurosci. Methods **284**, 27–34 (2017)
10. S. Paul, N. Kathmann, A. Riesel, The costs of distraction: the effect of distraction during repeated picture processing on the LPP. Biol. Psychol. **117**, 225–234 (2016)
11. M.A. Ghazanfar, M. Cook, B. Tang, I. Tait, A. Alijani, The effect of divided attention on novices and experts in laparoscopic task performance. Surg. Endosc. **29**(3), 614–619 (2015)
12. E. Vaportzis, N. Georgiou-Karistianis, J.C. Stout, Age and task difficulty differences in dual tasking using circle tracing and serial subtraction tasks. Aging Clin. Exp. Res. **26**(2), 201–211 (2014)
13. S.D. Newman, T.A. Keller, M.A. Just, Volitional control of attention and brain activation in dual task performance. Hum. Brain Mapp. **28**(2), 109–117 (2007)
14. Y. Liu, H. Ayaz, A. Curtin, P.A. Shewokis, B. Onaral, Detection of attention shift for asynchronous P300-based BCI. Conf. Proc. IEEE Eng. Med. Biol. Soc. **3850**, 3850–3853 (2012)
15. R.N. Roy, S. Bonnet, S. Charbonnier, A. Campagne, Mental fatigue and working memory load estimation: interaction and implications for EEG-based passive BCI. Conf. Proc. IEEE Eng. Med. Biol. Soc. **6607**, 6607–6610 (2013)
16. L.V. Kulke, J. Atkinson, O. Braddick, Neural differences between covert and overt attention studied using EEG with simultaneous remote eye tracking. Front. Hum. Neurosci. **10**, 592 (2016)
17. P. Praamstra, L. Boutsen, G.W. Humphreys, Frontoparietal control of spatial attention and motor intention in human EEG. J. Neurophysiol. **94**(1), 764–774 (2005)
18. P. Horki, G. Bauernfeind, W. Schippinger, G. Pichler, G.R. Müller-Putz, Evaluation of induced and evoked changes in EEG during selective attention to verbal stimuli. J. Neurosci. Methods **270**, 165–176 (2016)
19. J. Ramirez, M. Bomba, A. Singhal, B. Fowler, Influence of a visual spatial attention task on auditory early and late Nd and P300. Biol. Psychol. **68**(2), 121–134 (2005)
20. J.M. Sangal, R.B. Sangal, B. Persky, Abnormal auditory P300 topography in attention deficit disorder predicts poor response to pemoline. Clin. Electroencephalogr. **26**(4), 204–213 (1995)
21. S. Aliakbaryhosseinabadi, V. Kostic, A. Pavlovic, S. Radovanovic, E.N. Kamavuako, N. Jiang et al., Influence of attention alternation on movement-related cortical potentials in healthy individuals and stroke patients. Clin. Neurophysiology. **128**(1), 165–175 (2017)

22. A. Myrden, T. Chau, Effects of user mental state on EEG-BCI performance. Front. Hum. Neurosci. **9**, 308 (2015)
23. D.O. Hebb, *The organization of behavior: a neuropsychological theory* (L. Erlbaum Associates, Mahwah, N.J., 2002)
24. N. Mrachacz-Kersting, S.R. Kristensen, I.K. Niazi, D. Farina, Precise temporal association between cortical potentials evoked by motor imagination and afference induces cortical plasticity. J. Physiol. **590**(7), 1669–1682 (2012)
25. U. Ziemann, W. Paulus, M.A. Nitsche, A. Pascual-Leone, W.D. Byblow, A. Berardelli et al., Consensus: motor cortex plasticity protocols. Brain Stimul. **1**(3), 164–1682 (2008)
26. K. Stefan, M. Wycislo, J. Classen, Modulation of associative human motor cortical plasticity by attention. J. Neurophysiol. **92**(1), 66–72 (2004)
27. M. Pekna, M. Pekny, M. Nilsson, Modulation of neural plasticity as a basis for stroke rehabilitation. Stroke **43**(10), 2819–2828 (2012)
28. H. Cecotti, R.W. Kasper, J.C. Elliott, M.P. Eckstein, B. Giesbrecht, Multimodal target detection using single trial evoked EEG responses in single and dual-tasks. Conf. Proc. IEEE Eng. Med. Biol. Soc. **2011**, 6311–6314 (2011)
29. S.A. Bunge, T. Klingberg, R.B. Jacobsen, J.D. Gabrieli, A resource model of the neural basis of executive working memory. Proc. Natl. Acad. Sci. U S A **97**(7), 3573–3578 (2000)
30. F. Nijboer, E.W. Sellers, J. Mellinger, M.A. Jordan, T. Matuz, A. Furdea et al., A P300-based brain–computer interface for people with amyotrophic lateral sclerosis. Clin. Neurophysiol. **119**(8), 1909–1916 (2008)
31. J.W. Grabbe, P.A. Allen, Cross-task compatibility and age-related dual-task performance. Exp. Aging Res. **38**(5), 469–487 (2012)
32. E. Salo, T. Rinne, O. Salonen, K. Alho, Brain activity during auditory and visual phonological, spatial and simple discrimination tasks. Brain Res. **1496**, 55–69 (2013)
33. D.D. Salvucci, N.A. Taatgen, Threaded cognition: an integrated theory of concurrent multitasking. Psychol. Rev. **115**(1), 101–130 (2008)
34. K.M. Halvorson, What causes dual-task costs? University of Lowa (2013)
35. E. Kodesh, R. Kizony, Measuring cardiopulmonary parameters during dual-task while walking. J. Basic Clin. Physiol. Pharmacol. **25**(2), 155–160 (2014)
36. O. Beauchet, C. Annweiler, V. Dubost, G. Allali, R.W. Kressig, S. Bridenbaugh et al., Stops walking when talking: a predictor of falls in older adults? Eur. J. Neurol. **16**(7), 786–795 (2009)
37. T. Asai, T. Doi, S. Hirata, H. Ando, Dual tasking affects lateral trunk control in healthy younger and older adults. Gait Posture **38**(4), 830–836 (2013)
38. W.W.N. Tsang, N.K.Y. Lam, K.N.L. Lau, H.C.H. Leung, C.M.S. Tsang, X. Lu, The effects of aging on postural control and selective attention when stepping down while performing a concurrent auditory response task. Eur. J. Appl. Physiol. **113**(12), 3021–3026 (2013)
39. P.A. Allen, M. Sliwinski, T. Bowie, Differential age effects in semantic and episodic memory, Part II: slope and intercept analyses. Exp. Aging Res. **28**(2), 111–142 (2002)
40. P.A. Allen, E. Ruthruff, J.D. Elicker, M. Lien, Multisession, dual-task psychological refractory period practice benefits older and younger adults equally. Exp. Aging Res. **35**(4), 369,369–399 (2009)
41. V.L. Barbour, G.E. Mead, Fatigue after stroke: the patient's perspective. Stroke Res. Treat. **2012**, 1–6 (2012)
42. J. Serences, EEG and fMRI provide different insights into the link between attention and behavior in human visual cortex. J. Vision **15**(12), 1413 (2015)
43. Y. Fu, B. Xu, Y. Li, Y. Wang, Z. Yu, H. Li, Single-trial decoding of imagined grip force parameters involving the right or left hand based on movement-related cortical potentials. Chin. Sci. Bull. **59**(16), 1907–1916 (2014)
44. S.K. Kisiel, V. Siemionow, L.D. Zhang, J. Mencel, G.H. Yue, A. Jaskolska et al., The magnitude of the movement-related cortical potentials during fast and slow voluntary knee extensor deactivation. Acta Neurobiol. Exp. **72**(2), 204–205 (2012)
45. S. Aliakbaryhosseinabadi, N. Jiang, L. Petrini, D. Farina, K. Dremstrup, N. Mrachacz-Kersting, Robustness of movement detection techniques from motor execution: Single trial movement related cortical potential. Conf. Proc. IEEE NER, 2015 (2015)

46. R. Palaniappan, Brain computer interface design using band powers extracted during mental tasks. Conf. Proc. IEEE EMBS Neural. Eng, 2005 (2005)
47. A.S. Gevins, J.C. Doyle, G.M. Zeitlin, Comments on "the clinical EEG-a search for a buried message". IEEE Trans. Biomed. Eng. **25**(6), 568–569 (1978)
48. A.M. Waters, J. Henry, K. Mogg, B.P. Bradley, D.S. Pine, Attentional bias towards angry faces in childhood anxiety disorders. J. Behav. Ther. Exp. Psychiatry **41**(2), 158–164 (2010)
49. T.O. Zander, C. Kothe, Towards passive brain–computer interfaces: applying brain–computer interface technology to human–machine systems in general. J. Neural Eng. **8**(2), 025005 (2011)
50. K. Dijkstra, P. Brunner, A. Gunduz, W. Coon, A.L. Ritaccio, J. Farquhar et al., Identifying the attended speaker using electrocorticographic (ECoG) signals. Brain Comput. Interfaces **2**(4), 161–173 (2015)
51. R.V. Chacko, B. Kim, S.W. Jung, A.L. Daitch, J.L. Roland, N.V. Metcalf et al., Distinct phase-amplitude couplings distinguish cognitive processes in human attention. Neuroimage **175**, 111–121 (2018)
52. R.F. Helfrich, I.C. Fiebelkorn, S.M. Szczepanski, J.J. Lin, J. Parvizi, R.T. Knight et al., Neural mechanisms of sustained attention are rhythmic. Neuron **99**(4), 854–865 (2018)

Using a BCI Prosthetic Hand to Control Phantom Limb Pain

Takufumi Yanagisawa, Ryohei Fukuma, Ben Seymour, Koichi Hosomi, Haruhiko Kishima, Takeshi Shimizu, Hiroshi Yokoi, Masayuki Hirata, Toshiki Yoshimine, Yukiyasu Kamitani and Youichi Saitoh

Abstract Phantom limb pain is neuropathic pain that occurs after the amputation of a limb and partial or complete deafferentation. The underlying cause has been attributed to maladaptive plasticity of the sensorimotor cortex, and evidence suggests that experimental induction of further reorganization should affect the pain. Here, we use a brain–computer interface (BCI) based on real-time magnetoencephalography signals to reconstruct affected hand movements with a robotic hand. BCI training successfully induced some plastic alteration in the sensorimotor representation of the phantom hand movements. If a patient tried to control the robotic hand by associating the representation of phantom hand movement, it increased the pain while improving classification accuracy of the phantom hand movements. However, if the patient tried to control the robotic hand by associating the representation of the intact hand, it decreased the pain while decreasing the classification accuracy of the phantom hand movements. These results demonstrate that the BCI training controls the phantom limb pain depending on the induced sensorimotor plasticity. Moreover, these results strongly suggest that a reorganization of the sensorimotor cortex is the underlying cause of phantom limb pain.

T. Yanagisawa (✉) · R. Fukuma · K. Hosomi · H. Kishima · T. Shimizu · M. Hirata · T. Yoshimine · Y. Saitoh
Department of Neurosurgery, Osaka University Graduate School of Medicine, Osaka, Japan
e-mail: tyanagisawa@nsurg.med.osaka-u.ac.jp

T. Yanagisawa
Division of Functional Diagnostic Science, Osaka University Graduate School of Medicine, Osaka, Japan

T. Yanagisawa · R. Fukuma · Y. Kamitani
ATR Computational Neuroscience Laboratories, Department of Neuroinformatics, Kyoto, Japan

T. Yanagisawa · R. Fukuma · M. Hirata · T. Yoshimine
CiNet Computational Neuroscience Laboratories, Department of Neuroinformatics, Osaka, Japan

T. Yanagisawa
JST PRESTO, Osaka, Japan

Division of Clinical Neuroengineering, Global Center for Medical Engineering and Informatics, Osaka University, Osaka, Japan

© The Author(s), under exclusive licence to Springer Nature Switzerland AG 2019   43
C. Guger et al. (eds.), *Brain-Computer Interface Research*,
SpringerBriefs in Electrical and Computer Engineering,
https://doi.org/10.1007/978-3-030-05668-1_4

Keywords Cortical plasticity · Magnetoencephalography · Neurofeedback ·
Phantom limb pain · Robotic hand

1 Introduction

Phantom limb pain is an intractable chronic pain [1] that frequently occurs in a partially or completely deafferented body part after severe peripheral nerve injury [2] or amputation [3]. The pain has been attributed to maladaptive plasticity of the sensorimotor cortex [3–5]. Although previous studies reported correlations between pain and topographic reorganization of sensorimotor cortical maps [6–8], more recent studies provided conflicting evidence [9, 10] and questioned the maladaptive sensorimotor reorganization model, especially in terms of the cortical representation of phantom hand movements. To clarify the causative link between sensorimotor cortical plasticity and pain, direct experimental manipulation of sensorimotor plasticity is necessary.

A brain–computer interface (BCI) is a powerful tool that can induce plastic changes in cortical activities [11–13]. BCIs can be used to decode neural activity relating to hand movements and then convert the decoded movements into those of a prosthetic hand [14–20]. Hand movements can be precisely decoded using magnetoencephalography (MEG) signals [21, 22], even in severely paralyzed patients simply intending to move the affected hand [23, 24]. Moreover, BCI training induces plastic changes in cortical activity [25, 26] and potentially in associated clinical symptoms [27].

We applied BCI training of a prosthetic hand using real-time MEG signals to phantom limb patients and evaluated the association between changes in pain and in cortical activities relating to phantom hand movements [28]. It has been hypothesized

R. Fukuma · Y. Kamitani
Graduate School of Information Science, Nara Institute of Science and Technology, Nara, Japan

B. Seymour
Computational and Biological Learning Laboratory, Department of Engineering, University of Cambridge, Cambridge, UK

Center for Information and Neural Networks, National Institute for Information and Communications Technology, Osaka, Japan

K. Hosomi · T. Shimizu · Y. Saitoh
Department of Neuromodulation and Neurosurgery, Graduate School of Medicine, Osaka University, Osaka, Japan

H. Yokoi
Department of Mechanical Engineering and Intelligent Systems, The University of Electro-Communications, Tokyo, Japan

Y. Kamitani
Graduate School of Informatics, Kyoto University, Kyoto, Japan

that successful BCI training using decoded phantom hand movements should reduce pain with concurrent plastic changes in cortical activity. Although BCI training led to a significant increase of movement information in the sensorimotor cortex, the training significantly increased pain. In contrast, BCI training to associate a robotic hand with the intact hand representation reduced pain while decreasing classification accuracy of phantom hand movements. These results suggest a causative relationship between sensorimotor cortical plasticity and pain.

2 Subjects and Methods

2.1 Subjects

The study participants were nine brachial plexus root avulsion patients and one amputee (all males; mean age, 51.7 years; range, 38–60 years) who all had pain in their phantom limb (Table 1). The study adhered to the Declaration of Helsinki and was performed in accordance with protocols approved by the Ethics Committee of Osaka University Clinical Trial Center (No. 12107, UMIN000010180). All patients were informed of the purpose and possible consequences of this study, and written informed consent was obtained.

2.2 MEG Recording

Subjects were in the supine position with the head centered in the gantry. MEG signals were measured by a 160-channel whole-head MEG equipped with coaxial-type gradiometers housed in a magnetically shielded room (MEGvision NEO; Yokogawa Electric Corporation, Kanazawa, Japan). Signals were sampled at 1000 Hz with an online low-pass filter at 200 Hz and acquired online by FPGA DAQ boards (PXI-7854R; National Instruments, Austin, TX, USA) after passing through an optical isolation circuit. The signals for the 84 selected channels were used for offline analysis and online control of the prosthesis.

2.3 Experimental Design

All patients participated in a crossover trial consisting of three experiments on different days. Each experiment had an offline task (pre-BCI), BCI training, and an offline task (post-BCI) (Fig. 1). In the first offline task, patients attempted to move their phantom hands or their intact hands (grasping and opening) at the presented times [16] while the MEG signals of selected channels were recorded. After the offline

Table 1 Clinical profiles of patients

Patient ID	Age (years)/sex	Diagnosis	JART FSIQ/VIQ/PIQ	Disease duration (years)	Mirror therapy
1	50/M	Right BPRA of C6-8	100/100/90	34	Effective only for a short period
2	51/M	Left BPRA of C5-Th1	96/96/96	6	Not effective
3	58/M	Right BPRA of C6-Th2	112/114/108	40	No experience
4	48/M	Amputation below right elbow	104/105/102	1.5	Not effective
5	49/M	Right BPRA of C7-Th1	102/102/101	29	No experience
6	56/M	Right BPRA of C7-8	114/116/110	38	Not effective
7	51/M	Right BPRA of C6-Th1	110/112/107	11	No experience
8	56/M	Left BPRA of C7-Th1	83/82/87	13	Not effective
9	38/M	Right BPRA of C6-8	85/84/89	21	No experience
10	60/M	Right BPRA of C6-8	114/116/110	20	No experience

M male; *BPRA* brachial plexus root avulsion; *JART* Japanese Adult Reading Test; *FSIQ* full-scale intelligence quotient; *VIQ* verbal intelligence quotient; *PIQ* performance intelligence quotient

task, pain was evaluated with a visual analog scale (VAS). The acquired MEG signals were used to construct the decoder to control the robotic hand [24]. Then, the subjects were instructed to control the robotic hand in real time using the trained decoder. They were instructed to freely control the robotic hand to grasp and release a ball by trying to move their phantom hands while watching the movement of the prosthetic hand in closed-loop conditions. The robotic hand was controlled according to the movements inferred by a selected decoder with the MEG sensor signals obtained online [24].

The experiment was performed three times with different decoders, with an interval of at least 2 weeks between experiments. The order of the phantom decoder and random decoder experiments was randomly assigned to the patients to balance group sizes, and then the experiment with the real hand decoder was performed.

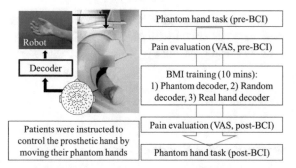

Fig. 1 BCI training and experimental design. (Left) Patients were instructed to control the robot while watching the robot's movement through the monitor. The robot hand was controlled by the decoder based on the MEG signals acquired online. **(Right)** The tasks in each experiment. During the 10 min of BCI training, three types of decoders were used to control the robotic hand, each for three experiments. For the experiment with the real hand decoder, the patients also performed the offline task with their intact hand after the task with their phantom hand

2.4 Cortical Current Estimation by VBMEG

A polygonal model of the cortical surface was constructed based on structural MRI (T1-weighted; Signa HDxt Excite 3.0T; GE Healthcare UK Ltd., Buckinghamshire, UK) using the Freesurfer software (http://surfer.nmr.mgh.harvard.edu/) [29]. To align MEG data with individual MRI data, we scanned the three-dimensional facial surface and 50 points on the scalp of each participant (FastSCAN Cobra; Polhemus, Colchester, VT, USA). Three-dimensional facial surface data were superimposed on the anatomical facial surface provided by the MRI data. The positions of five marker coils before each recording were used to estimate cortical current with VBMEG (ATR Neural Information Analysis Laboratories, Kyoto, Japan) [30, 31]. The hyper-parameters m0 and γ0 were set to 100 and 10, respectively. The inverse filter was estimated by using MEG signals in all trials from 0 to 1 s in the offline task, with the baseline of the current variance estimated from the signals from -1.5 to -0.5 s. The filter was then applied to sensor signals in each trial to calculate cortical currents.

2.5 F-Values and Classification of Movement Types in the Offline Task

The estimated cortical currents in the sensorimotor cortex were converted to z-scores by using the mean and standard deviation of the currents. Then, the z-scored cortical currents were compared between two types of movements with a one-way analysis of variance (ANOVA) for each vertex. The F-value of the ANOVA was plotted at each vertex. Movement types were also classified using the z-scored cortical currents. We trained the decoder by using a support vector machine [24]. Classification accuracy of the movement type was estimated by 10-fold nested cross-validation [21]. All decoding analyses were performed with MATLAB R2013a using the radial basis function kernel support vector machine.

3 Results

3.1 BCI Training with a Robotic Hand

Figure 2 shows the F-values of the z-scored cortical currents at the execution cue for Patient 4. After BCI training with the phantom decoder, the F-values increased in the contralateral sensorimotor cortex (Fig. 2a), but the pain score also increased from 52 to 57 on the VAS (a scale of 1–100). However, after BCI training with the real hand decoder, the F-values decreased in the contralateral sensorimotor cortex (Fig. 2b), and the pain score decreased from 42 to 40.

3.2 Pain and Classification Accuracy

Among the 10 patients, increases in the pain VAS scores were significantly changed depending on the decoder type ($n = 10$ each, $p = 0.0002$, $F(2, 27) = 11.5$, one-way ANOVA) (Fig. 3a). After training with the real hand decoder, the VAS scores decreased significantly compared to those of the random decoder and the phantom decoder ($n = 10$, $p = 0.025$ and 0.0003, uncorrected, $t(18) = 2.45$ and 4.36, respectively, two-tailed Student t-test). In contrast, the VAS scores increased significantly

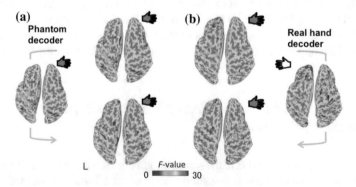

Fig. 2 Sensorimotor plasticity induced by BCI training. Significant F-values for the two phantom movements at each vertex were color-coded on Patient 4's brain surfaces for the pre-training (upper) and post-training (lower). The image to the side shows the F-value of data used for the decoder construction. **a** Phantom hand decoder experiment; **b** real hand decoder experiment

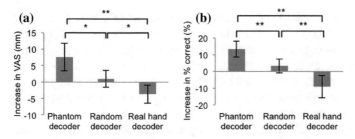

Fig. 3 Alteration in pain and classification accuracies among the three experiments. a The averaged differences in VAS scores (post–pre) are shown with the 95% confidence interval for three experiments ($n = 10$; *$p < 0.05$, **$p < 0.01$, uncorrected, two-tailed Student t-test). **b** The accuracies for classifying two types of phantom movements were evaluated using the currents on the motor cortex contralateral to the phantom hand. Each bar shows the average difference in the accuracy with 95% confidence intervals for each experiment ($n = 10$, **$p < 0.01$, Bonferroni-corrected, two-tailed Student t-test)

after training with the phantom decoder compared to the random decoder ($n = 10$, $p = 0.017$, uncorrected, $t(18) = 2.62$, two-tailed Student t-test). Notably, these increased scores from the phantom decoder spontaneously returned to the previous state after more than 2 weeks and were not significantly different from pre-training scores ($n = 10$, $p = 0.55$, uncorrected, $t(9) = 0.63$, paired Student t-test).

We also compared the accuracies for classifying the phantom movements using the estimated currents in the sensorimotor cortex contralateral to the phantom hand. The differences in accuracies varied significantly among the three conditions using the currents contralateral to the phantom hand ($n = 10$ each, $p = 0.00001$, $F(2, 27) = 17.52$, one-way ANOVA) (Fig. 3b). The classification accuracy decreased significantly after training with the real hand decoder compared to the phantom decoder and random decoder ($n = 10$, $p = 0.00004$ and 0.006, Bonferroni-corrected, $t(18) = 7.44$ and 3.59, respectively, two-tailed Student t-test), whereas it increased significantly after training with the phantom decoder compared to the random decoder ($n = 10$, $p = 0.006$, Bonferroni-corrected, $t(18) = 3.60$, two-tailed Student t-test).

Changes in the VAS scores were significantly correlated with changes in classification accuracy using the currents for the sensorimotor cortex contralateral to the phantom hand ($n = 30$, $R = 0.66$, $p = 0.0001$, Pearson correlation coefficient), but they were not significantly correlated with the changes in accuracy using the currents ipsilateral to the phantom hand ($n = 30$, $R = 0.037$, $p = 0.76$).

4 Discussion

Our data demonstrated that MEG-based BCI training to control a robotic hand significantly changed the pain and classification accuracy using the contralateral sensorimotor cortical currents. Pain increased significantly in proportion to the classification accuracy of phantom hand movements. Subsequent training of phantom hand movements based on the intact hand disrupted the information content of the contralateral sensorimotor cortex and led to an improvement in pain. These results are consistent with a causal relationship between cortical sensorimotor plasticity and phantom limb pain. By selecting the decoded information to control the BCI, we induced plasticity in the cortical information representation of phantom hand movements to explore the relationship with pain [32].

As expected, BCI training with the phantom decoder increased the classification accuracy of the phantom hand movements using the cortical currents contralateral to the phantom hand. However, during training with the real hand decoder, the patients intended to associate their phantom hand movements with movements of the prosthetic hand, which was actually controlled by a decoder to classify the MEG signals based on the intact hand's movement. As a result, the patients were expected to unknowingly associate the phantom hand movements with the cortical representation of the intact hand's movements, which were different from the cortical representation of the phantom movements in pre-BCI training. BCI training with the real hand decoder would accelerate the dissociation of the link between the phantom hand and the original cortical representation by creating a new link to the real hand. The association of the different neural representation might dissociate the prosthetic hand and the original neural representation of phantom movements even more so than the association of the randomly moved prosthetic hand and the neural representation.

Recent studies demonstrated that chronic pain is a maladaptive neurological disease state [33, 34]. To understand the mechanism of pain, we need to understand how the information relating to pain is processed in the brain. Moreover, to control chronic pain, we need to control the information processing relating to the pain. Our results demonstrated that BCI training is a useful interventional tool that can be used to study and control information processing in the brain. BCI training induces functionally specific plastic changes in the targeted cortex, based on the information represented in the activity. In addition, this method can be easily applied to patients in non-invasive, randomized, and blinded studies. By combining decoding [35] and neurofeedback, BCI could be applied to treat chronic pain.

In addition, the results clearly indicate that we should consider pain as a potential complication when using BCIs for paralyzed patients. That is, we must consider the fact that true decoding that increases pain creates a problem for those patients using BCI-controlled robotic prostheses. One possible solution, according to the incongruence hypothesis, would be BCI training accompanied by sensory feedback (e.g., artificial–real nerve coupling [36]), which might mitigate pain by providing an intact sensorimotor loop.

In summary, neurofeedback training using MEG-based BCI provides a novel method to directly change the information content of motor representations by inducing plasticity in the sensorimotor cortex. Our experiments showed that BCI training to enhance phantom limb representation was associated with increased pain, and BCI training to deteriorate the representation reduced pain. This suggests a direct and causative link between sensorimotor cortical plasticity and pain in phantom limb patients, and that BCI training may be a novel and clinically useful treatment.

Acknowledgements This research was conducted under the "Development of BMI Technologies for Clinical Application" of SRPBS by MEXT and AMED. This research was also supported in part by JST PRESTO; Grants-in-Aid for Scientific Research KAKENHI (90533802, 24700419, 26560467, 26242088, 22700435, 17H06032, 15H05710, 15H05920); Brain/MINDS and SICP from AMED; ImPACT; Ministry of Health, Labor, and Welfare (18261201); the Japan Foundation of Aging and Health and TERUMO foundation for life sciences and arts.

References

1. A. Wolff et al., 21. Phantom pain. Pain Pract **11**(4), 403–413 (2011)
2. H. Shankar, J. Hansen, K. Thomas, Phantom pain in a patient with brachial plexus avulsion injury. Pain Med. **16**(4), 777–781 (2015)
3. H. Flor, L. Nikolajsen, T. Staehelin Jensen, Phantom limb pain: A case of maladaptive CNS plasticity? Nat. Rev. Neurosci. 7(11), 873–881 (2006)
4. H. Flor, N. Birbaumer, Phantom limb pain: cortical plasticity and novel therapeutic approaches. Curr. Opin. Anaesthesiol. **13**(5), 561–564 (2000)
5. V.S. Ramachandran, D. Rogers-Ramachandran, S. Cobb, Touching the phantom limb. Nature **377**(6549), 489–490 (1995)
6. H. Flor et al., Phantom-limb pain as a perceptual correlate of cortical reorganization following arm amputation. Nature **375**(6531), 482–484 (1995)
7. M. Lotze et al., Phantom movements and pain: an fMRI study in upper limb amputees. Brain **124**(11), 2268–2277 (2001)
8. A. Karl et al., Reorganization of motor and somatosensory cortex in upper extremity amputees with phantom limb pain. J. Neurosci. **21**(10), 3609–3618 (2001)
9. T.R. Makin et al., Reassessing cortical reorganization in the primary sensorimotor cortex following arm amputation. Brain **138**(8), 2140–2146 (2015)
10. T.R. Makin et al., Phantom pain is associated with preserved structure and function in the former hand area. Nat. Commun. **4**, 1570 (2013)
11. A.L. Orsborn et al., Closed-loop decoder adaptation shapes neural plasticity for skillful neuroprosthetic control. Neuron **82**(6), 1380–1393 (2014)
12. K. Ganguly et al., Reversible large-scale modification of cortical networks during neuroprosthetic control. Nat. Neurosci. **14**(5), 662–667 (2011)
13. J.D. Wander et al., Distributed cortical adaptation during learning of a brain-computer interface task. Proc. Natl. Acad. Sci. U.S.A. **110**(26), 10818–10823 (2013)
14. Y. Nakanishi et al., Decoding fingertip trajectory from electrocorticographic signals in humans. Neurosci. Res. **85**, 20–27 (2014)
15. Y. Nakanishi et al., Prediction of three-dimensional arm trajectories based on ECoG signals recorded from human sensorimotor cortex. PLoS ONE **8**(8), e72085 (2013)
16. T. Yanagisawa et al., Electrocorticographic control of a prosthetic arm in paralyzed patients. Ann. Neurol. **71**(3), 353–361 (2012)
17. T. Yanagisawa et al., Real-time control of a prosthetic hand using human electrocorticography signals. J. Neurosurg. **114**(6), 1715–1722 (2011)
18. T. Yanagisawa et al., Neural decoding using gyral and intrasulcal electrocorticograms. Neuroimage **45**(4), 1099–1106 (2009)
19. T. Yanagisawa et al., Regulation of motor representation by phase-amplitude coupling in the sensorimotor cortex. J. Neurosci. **32**(44), 15467–15475 (2012)
20. T. Yanagisawa et al., Movement induces suppression of interictal spikes in sensorimotor neocortical epilepsy. Epilepsy Res. **87**(1), 12–17 (2009)
21. R. Fukuma et al., Closed-loop control of a neuroprosthetic hand by magnetoencephalographic signals. PLoS ONE **10**(7), e0131547 (2015)

22. A. Toda et al., Reconstruction of two-dimensional movement trajectories from selected magnetoencephalography cortical currents by combined sparse Bayesian methods. NeuroImage **54**(2), 892–905 (2011)
23. E. Buch et al., Think to move: a neuromagnetic brain-computer interface (BCI) system for chronic stroke. Stroke **39**(3), 910–917 (2008)
24. R. Fukuma et al., Real-time control of a neuroprosthetic hand by magnetoencephalographic signals from paralysed patients. Sci. Rep. **6**, 21781 (2016)
25. Y. Nishimura et al., Spike-timing-dependent plasticity in primate corticospinal connections induced during free behavior. Neuron **80**(5), 1301–1309 (2013)
26. K.B. Clancy et al., Volitional modulation of optically recorded calcium signals during neuroprosthetic learning. Nat. Neurosci. **17**(6), 807–809 (2014)
27. E.R. Buch et al., Parietofrontal integrity determines neural modulation associated with grasping imagery after stroke. Brain **135**(2), 596–614 (2012)
28. T. Yanagisawa et al., Induced sensorimotor brain plasticity controls pain in phantom limb patients. Nat. Commun. **7**, 13209 (2016)
29. A.M. Dale, B. Fischl, M.I. Sereno, Cortical surface-based analysis. I. Segmentation and surface reconstruction. NeuroImage **9**(2), 179–194 (1999)
30. L.G. Cohen et al., Motor reorganization after upper limb amputation in man: a study with focal magnetic stimulation. Brain **114**(1B), 615–627 (1991)
31. T. Yoshioka et al., Evaluation of hierarchical Bayesian method through retinotopic brain activities reconstruction from fMRI and MEG signals. Neuroimage **42**(4), 1397–1413 (2008)
32. K. Shibata et al., Perceptual learning incepted by decoded fMRI neurofeedback without stimulus presentation. Science **334**(6061), 1413–1415 (2011)
33. M.N. Baliki, A.V. Apkarian, Nociception, pain, negative moods, and behavior selection. Neuron **87**(3), 474–491 (2015)
34. R. Kuner, H. Flor, Structural plasticity and reorganisation in chronic pain. Nat. Rev. Neurosci. **18**(1), 20–30 (2016)
35. T.D. Wager et al., An fMRI-based neurologic signature of physical pain. N. Engl. J. Med. **368**(15), 1388–1397 (2013)
36. S. Raspopovic et al., Restoring natural sensory feedback in real-time bidirectional hand prostheses. Sci. Transl. Med. **6**(222), 222ra19 (2014)

Restoration of Finger and Arm Movements Using Hybrid Brain/Neural Assistive Technology in Everyday Life Environments

Surjo R. Soekadar, Marius Nann, Simona Crea, Emilio Trigili,
Cristina Gómez, Eloy Opisso, Leonardo G. Cohen, Niels Birbaumer
and Nicola Vitiello

Abstract Controlling advanced robotic systems with brain signals promises substantial improvements in health care, for example, to restore intuitive control of hand movements after severe stroke or spinal cord injuries (SCI). However, such integrated, brain- or neural-controlled robotic systems have yet to enter broader clinical use or daily life environments. The main challenge to integrate such systems in everyday life environments relates to the reliability of brain-control, particularly when brain signals are recorded non-invasively. Using a non-invasive, hybrid EEG-EOG-based brain/neural hand exoskeleton (B/NHE), we demonstrate full restoration of activities of daily living (ADL), such as eating and drinking, across six paraplegic individuals (five males, 30 ± 14 years) outside the laboratory. In a second set of experiments, we show that even whole-arm exoskeleton control is feasible and safe by combining hybrid brain/neural control with vision-guided and context-sensitive autonomous robotics. Given that recent studies indicate neurological recovery after chronic stroke or SCI when brain-controlled assistive technology is repeatedly used

S. R. Soekadar (✉) · M. Nann
Clinical Neurotechnology Lab, Neuroscience Research Center (NWFZ), Department of
Psychiatry and Psychotherapy, Charité - University Medicine Berlin, Berlin, Germany
e-mail: surjo@soekadar.com

Applied Neurotechnology Lab, Department of Psychiatry and Psychotherapy, University Hospital
of Tübingen, Tübingen, Germany

S. Crea · E. Trigili · N. Vitiello
The BioRobotics Institute, Scuola Superiore Sant'Anna, Pisa, Italy

C. Gómez · E. Opisso
Hospital de Neurorehabilitació Institut Guttmann, Barcelona, Spain

L. G. Cohen
Human Cortical Physiology and Stroke Neurorehabilitation Section, National Institutes of
Neurological Disorders and Stroke, National Institutes of Health, Bethesda, USA

N. Birbaumer
Institute of Medical Psychology and Behavioral Neurobiology, Eberhard Karls University of
Tübingen, Tübingen, Germany

Wyss Center for Bio and Neuroengineering, Geneva, Switzerland

© The Author(s), under exclusive licence to Springer Nature Switzerland AG 2019   53
C. Guger et al. (eds.), *Brain-Computer Interface Research*,
SpringerBriefs in Electrical and Computer Engineering,
https://doi.org/10.1007/978-3-030-05668-1_5

for 1–12 months, we suggest that combining an *assistive* and *rehabilitative* approach may further promote brain-machine interface (BMI) technology as a standard therapy option after stroke and SCI. In such scenario, brain/neural-assistive technology would not only have an immediate impact on the quality of life and autonomy of individuals with brain or spinal cord lesions but would also foster neurological recovery by stimulating functional and structural neuroplasticity.

Keywords Brain/neural hand exoskeleton (B/NHE) · Spinal cord injury (SCI) · Stroke · Activities of daily living (ADL) · Neural recovery

1 Introduction

Stroke and spinal cord injuries (SCI) are among the leading causes of finger paralysis resulting in long-term disability worldwide [24]. Over the coming years, the number of stroke and SCI survivors who depend on assistance in their everyday life will substantially increase due to demographic factors [2]. Approximately one third of all stroke survivors and all SCI survivors with high cervical lesions suffer from severe finger paralysis. For these patients, there is currently no standardized and accepted treatment strategy [18–20]. While SCI survivors with remaining wrist movements can undergo a tendon transfer surgery, such procedure restores hand function to a limited degree only and at the expense of other motor functions [14]. The social and economic burden of chronic finger paralysis after stroke or SCI is, thus, very high [12].

It was shown that rehabilitation-training that builds on existing motor functions of the affected hand while constraining use of compensatory strategies, e.g. using the less affected hand after stroke, can lead to substantial and sustained motor recovery [26]. However, such constrained induced movement therapy (CIMT) requires a sufficient degree of rest motor function that is absent in many stroke and SCI survivors.

The development of brain–computer or brain–machine interfaces (BCIs/BMIs) translating electric, magnetic or metabolic brain activity into control signals of external machines or robots has recently raised hopes that disrupted motor pathways could be bypassed and allow individuals with stroke or SCI to grasp using a brain-controlled robotic device, exoskeleton or functional electric stimulation (FES) of hand and finger muscles. By substituting for lost motor functions, such assistive BMIs have recently demonstrated recovery of versatile motor control [9, 13]. However, versatile brain-control of individual finger movements required implantation of microelectrodes, a procedure involving the risk of infections or bleeding.

Despite considerable efforts, e.g. implementation of intelligent machine learning algorithms or remarkable technical advances improving active BMI control, online classification of *non-invasive* recordings on the other side lacks reliability, a serious limitation in terms of applicability and safety in everyday life environments. A viable strategy to increase the safety of brain/neural assistive technology in everyday life environments is to use a switch mechanism that turns the BMI system off when active brain/neural control is not needed. Additionally, recent studies combined different biosignals, e.g. EEG and EOG signals, to increase the degrees of freedom in control of external devices, e.g. to navigate a wheelchair [22]. Given that eye-muscle control remains largely intact, even in severe neurodegenerative disorders, we reasoned that combining EEG and EOG might be a suitable strategy to control brain/neural assistive technology, e.g. a hand exoskeleton [18], after brain and spinal cord lesions.

2   Hybrid EEG/EOG Hand-Exoskeleton Control

In a first feasibility study, we investigated whether fusion of EEG and EOG signals can enhance reliability and safety of continuous brain control of a hand exoskeleton performing grasping motions [25]. In this study, 12 healthy volunteers (8 male, 4 female, mean age: 28.1 ± 3.63 years), who have never used a brain/neural hand exoskeleton before, participated in a 1-hour experimental session. After calibration of the system, brain/neural control and safety were evaluated by calculating motions in % relative to a full hand exosksleton closing during visual instruction to either close or not to close the device (indicated as a green and red square, respectively, shown for 5 s). A brain/neural-controlled grasping motion was defined as successful if the hand exoskeleton closed more than 50% while the green square was displayed. This value was chosen under the assumption that the size of most everyday life objects is smaller than 50% of the user's full hand span. Control of the device was defined as successful if average hand-closing motions during green square presentations exceeded more than 60% of full closing motions. Closing motions that exceeded 25% of a full hand closing while a red square was displayed were defined as safety violations. This value was chosen under the assumption that most daily life objects that are grasped are smaller than 75% of a full hand span.

We found that all participants were able to control the brain/neural hand exoskeleton using EEG and EOG signals. Using hybrid EEG/EOG control, the exoskeleton was closed in average by 60.77 ± 9.42% during green square presentations, and 12.31 ± 5.39% during red square presentations. By adding EOG control, safety violations occurred only in 10.14 ± 0.3% of red square presentations. The maximum closing motion during a red square presentation stopped at 28% of a full closing motion. These results confirmed that it is feasible and safe to use a hybrid EEG/EOG control paradigm to drive a hand exoskeleton [18, 25].

As a next step, we implemented this control paradigm into a portable and wireless system integrated into a wheelchair and tested the system in six quadriplegic patients with severe finger paralysis ([21], Fig. 1). The system translated electric brain activity in the mu-band (9–11 Hz) associated with the attempt to grasp into contingent closing motions of a hand exoskeleton attached to the participants' fingers. Hand opening is controlled by electric activity related to horizontal eye movements (HOV). While hand exoskeleton motions were set up to palmar grasp for large objects (>3 cm), motion trajectory changed to lateral pinch for smaller objects (<3 cm). This allowed versatile use of the system restoring almost 90% of normal hand function as measured

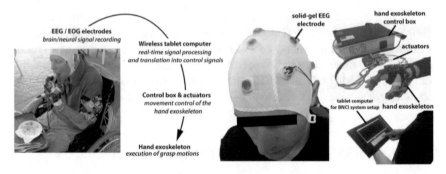

Fig. 1 Left panel: Process pipeline for hybrid control of a hand exoskeleton using electroencephalography and electrooculography (EEG/EOG) as used in [21]. EEG and EOG signals were sent to a tablet computer that processed and translated the recorded bio-signals into hand exoskeleton-driven grasping motions. Right panel: The system consisted of a solid-gel EEG system, a control box with motor units integrated into the study participants' wheelchair and a tablet computer to set up the BMI software (From Ref. [21], *Science Robotics*)

by the Toronto-Rehabilitation Hand Function Test (TRI-HFT). The BMI control software was operated through a tablet computer.

A major challenge in using electric brain signals for BMI control relates to the electrodes. Conventional wet-electrodes based on electrolyte paste require hair washing after the end of the session and dry- electrodes are often intolerable for long-term use (>1 h). Hence, solid-gel electrodes [22] were used, reducing preparation effort and making hair washing after use unnecessary. Integration of the system components into the study participants' wheelchair allowed users to leave the laboratory and operate the system independently in their everyday life environment, e.g. to eat and drink in an outside restaurant (Fig. 1).[1]

Besides the evident success of this approach, there are also some limitations related to the use of hand exoskeletons for motor restoration. These include the added weight that is put on the user's limb, possible restrictions in the upper limbs degrees of freedom (DOFs), but also component fatigue and natural wear of the working parts (as in all mechanical devices). The methodology's overall clinical validity, including in finger paralysis due to other causes such as stroke, traumatic brain injury (TBI) or neurodegenerative disorders, needs to be tested in larger clinical trials. Also, it will be critical to further facilitate mounting and unmounting of the system components, preferably to a degree at which no third-person assistance is needed.

As only grasp motions were restored, individuals without sufficient shoulder and arm motor function that limit their ability to reach out may not benefit from a hand exoskeleton. In this case, a whole-arm exoskeleton would be required. Due to the high number in degrees of freedom (DOF) (related to the shoulder, elbow and wrist joints), at present, reliable control cannot be achieved using a non-invasive BMI alone. For example, a seven degrees-of-freedom (DOF) system like the human arm,

---

[1]https://www.youtube.com/watch?v=zs5k7MpS1g0.

in which each joint can assume four discrete joint positions, would result in an actions space with a dimensionality of 4^7 (16.384). In contrast to a simple grasping task, operating a whole-arm exoskeleton, for example to drink, involves a series of sub-tasks such as reaching, grasping and lifting. Currently, information transfer rates (ITR) required for such high-dimensional control of a robotic whole-arm exoskeleton exceeds ITRs of any established non-invasive brain-machine interface (BMI) system. We thus integrated a hybrid EEG/EOG BMI with vision-guided autonomous robotics and tested whether such a novel paradigm can be used for fluent, reliable and safe whole-arm exoskeleton control.

3 Combining Hybrid EEG/EOG Control with Vision-Guided and Context-Sensitive Autonomous Robotics

Various neurological disorders related to brain lesions (e.g., stroke, TBI), neurodegeneration (e.g., amyotrophic lateral sclerosis, ALS) or neuroinflammation (e.g., multiple sclerosis, MS) can lead to complete loss of hand and arm function. To restore this function, we developed a whole-arm exoskeleton and tested whether the shared human-robot control strategy based on EEG/EOG signals described in Soekadar et al. [21] can be used for its fluent and reliable operation, e.g. to reach out, lift and drink from a cup [8]. Seven abled-bodied participants (seven right-handed males, mean age 30 ± 8 years) were instructed to use a hybrid EEG/EOG-controlled whole-arm exoskeleton that was attached to their right arm. Participants were asked to perform a drinking task comprising multiple sub-tasks (reaching, grasping, drinking, moving back and releasing a cup). Fluent and reliable control was defined as average 'time to initialize' (TTI) execution of each sub-task below 3 s with successful initializations of at least 75% of sub-tasks within 5 s.

We found that all participants were able to fluently and reliably control the vision-guided autonomous whole-arm exoskeleton (average TTI 2.12 ± 0.78 s across modalities with 75% successful initializations reached at 1.9 s for EOG and 4.1 s for EEG control). None of the participants complained about any discomfort or undesirable effects.

Further clinical studies are now needed to investigate whether these findings can be generalized to various clinical populations. Before such whole-arm exoskeletons can enter everyday life environments, however, a number of technical challenges need to be addressed. These challenges relate to the weight and mobility of the device, but also to its adaptability to the specific anatomical characteristics of the end-users. Another challenge relates to the safety of such systems: while our study was performed in a controlled lab environment, further studies are needed to investigate possible safety constraints when used in everyday life environments. In this respect, implementation of a reliable *veto function*, i.e. the ability to interrupt unintended motions or behaviors, will be critical [7]. Once addressed, however, shared EEG/EOG and vision-guided autonomous whole-arm exoskeletons promise to fully restore lost autonomy, even in complete arm and shoulder paralysis.

Besides the apparent and immediate restoration of autonomy and quality of life, there is another feature of brain-controlled devices that may have important implications for neurorestoration: It was shown that repeated use of brain-controlled exoskeletons can trigger neurological recovery [11, 15], a finding that may be of significance beyond the restoration of movement.

4   Repeated Use of Brain-Controlled Exoskeletons Triggers Neurological Recovery

First studies involving a non-invasive brain-controlled exoskeleton indicated that the majority of chronic stroke patients can learn to control such device using ipsilesional sensorimotor rhythms (SMR) [4, 17]. However, a few weeks of daily use did not result in any significant motor function improvement. When coupled with goal-directed behavioral physical therapy, however, clear improvements of motor and cognitive capacities were found [3]. These improvements were associated with increased activation of the ipsilesional hemisphere as well as increased fractional anisotropy in the ipsilesional corticospinal tract [5]. Motivated by these studies, a larger placebo-controlled clinical trial further corroborated this finding [15]. While the underlying mechanisms of the reported clinical improvements are not entirely understood, a possible mechanism may involve re-wiring and unmasking of remaining cortico-spinal fibers of the ipsilesional sensorimotor loop [1, 10, 18–20]. In this context, modulation of late-cortical disinhibition (LCD) may have played an important role [16]. A recent meta-analysis including data of 235 post-stroke survivors who participated in nine clinical trials showed that BMI use-dependent motor improvements, mostly quantified by the upper limb Fugl-Meyer Assessment (FMA-UE) score, exceeded the minimal clinically important difference (MCID) in six trials, while such improvement was only reached in three of nine control groups. In addition, several of the included studies indicated BMI-induced functional and structural neuroplasticity at a subclinical level [6].

Despite these very promising results, more and larger clinical trials are urgently needed, particularly investigating the underlying mechanisms of BMI-related motor recovery and to address other questions related to BMI training, for instance, dose-response relationships or the influence of BMI training on other brain functions, e.g. the cognitive domain [19].

Complementary to these studies involving stroke survivors, there is also some evidence that repeated use of brain-controlled exoskeletons can trigger neurological recovery in SCI survivors. In a first longitudinal study, eight chronic (3–13 years)

SCI survivors with quadriplegia participated in a long-term training based on a multi-stage BMI-based gait neurorehabilitation paradigm [11]. This paradigm combined immersive virtual reality (VR) training, visual-tactile feedback and EEG-triggered control of robotic actuators restoring walking motions. After one year of training, all eight participants showed neurological improvements in somatic sensation (pain localization, touch and proprioceptive sensing) over multiple dermatomes. Moreover, the participants also regained voluntary motor control in key muscles below the SCI level. As a result, 50% of the participants were clinically re-classified from complete to incomplete paraplegia.

A major limitation constraining broader clinical use of BMI-based neurorehabilitation relates to the costs required to train stroke survivors on a daily basis. This stems from a variety of reasons. First, study participants whose mobility is often severely compromised have to visit the clinical center or research facility on a daily basis. Hence, many decide to stay in nearby guest houses or ask for clinical admission, which increases the expenses for such studies. The studies are often performed in neurophysiological laboratories that need to be maintained. Moreover, current BMI systems and exoskeletons can be only set up and operated by well-trained and experienced personnel. Particularly, transfer of learned skills from the laboratory to everyday life activities requires a physiotherapist who assists and guides the patient over many weeks. Furthermore, the equipment currently used for BMI studies is rather expensive and there are no hand exoskeletons for patients with hand paralysis commercially available.

We thus suggest combining both approaches, the *assistive* and *restorative/rehabilitative use* of brain/neural-machine interfaces.

In a first prototype, we established a portable, lightweight and user-friendly brain/neural hand exoskeleton that was used in the patient's daily life environment (Fig. 2). The system comprised a 5-channel wireless headset EEG system. The hand exoskeleton was manufactured from individually-tailored 3D-printed, elastic parts that were lightweight, robust and washable. Electric motors actuated the fingers. The control software was implemented on a portable tablet computer that users can easily operate without any extensive familiarization.

In a follow-up prototype currently under development in a project supported by the Baden-Württemberg Foundation, a novel EEG headset will be used that hemiplegic patients can mount without any assistance. The system will track and record all relevant parameters and neurophysiological data during daily use of the system allowing for a therapist or physician to remotely supervise and optimize the daily training sessions. Availability of such system promises to further promote BMI technology as a standard therapy option after stroke or SCI. Besides having an immediate impact on the patients' autonomy and quality of life, thus providing an incentive for frequent use, the system may also possibly trigger neurological recovery. Availability of large neurophysiological and behavioral datasets as collected by the system will allow to elucidate the underlying mechanisms of BMI-related motor recovery, a critical prerequisite to extend the use of BMI technology to other domains of brain function, such as cognition, memory or emotion regulation.

Fig. 2 First prototype of a hybrid EEG/EOG-based brain/neural hand exoskeleton for *assistive* and *rehabilitative* use in everyday life environments. Besides immediate restoration of the ability to grasp and manipulate different objects of daily living, repeated use of the device may trigger neurological recovery

Acknowledgements This chapter and the presented studies were supported by the European Commission under the project AIDE (G.A. no: 645322), the European Research Council (ERC) under the project NGBMI (759370), and the Baden-Württemberg Stiftung (NEU007/1). SRS received special support by the Brain & Behavior Research Foundation as 2017 NARSAD Young Investigator Grant recipient and P&S Fund Investigator.

References

1. N. Birbaumer, L.G. Cohen, Brain-computer interfaces: communication and restoration of movement in paralysis. J. Physiol. **579**, 621–636 (2007)
2. G.L. Birbeck, M.G. Hanna, R.C. Griggs, Global opportunities and challenges for clinical neuroscience. JAMA **311**, 1609–1610 (2014)
3. D. Broetz, C. Braun, C. Weber, S.R. Soekadar, A. Caria, N. Birbaumer, Combination of brain-computer interface training and goal-directed physical therapy in chronic stroke: a case report. Neurorehabilitation and Neural Repair **24**, 674–679 (2010)
4. E. Buch, C. Weber, L.G. Cohen, C. Braun, M.A. Dimyan, T. Ard, J. Mellinger, A. Caria, S.R. Soekadar, A. Fourkas, N. Birbaumer, Think to move: a neuromagnetic brain-computer interface (BCI) system for chronic stroke. Stroke **39**, 910–917 (2008)
5. A. Caria, C. Weber, D. Brötz, A. Ramos, L.F. Ticini, A. Gharabaghi, C. Braun, N. Birbaumer, Chronic stroke recovery after combined BCI training and physiotherapy: a case report. J. Psychophysiol. **48**, 578–582 (2011)
6. M.A. Cervera, S.R. Soekadar, J. Ushiba, J.D.R. Millán, M. Liu, N. Birbaumer, G. Garipelli, Brain-computer interfaces for post-stroke motor rehabilitation: a meta-analysis. Ann. Clin. Transl. Neurol. **5**, 651–663 (2018)
7. J. Clausen, E. Fetz, J. Donoghue, J. Ushiba, U. Spörhase, J. Chandler, N. Birbaumer, S.R. Soekadar, Help, hope and hype: ethical dimensions of neuroprosthetics. Science **356**, 1338–1339 (2017)
8. S. Crea, M. Nann, E. Trigili, F. Cordella, A. Baldoni, F.J. Badesa, J.M. Catalan, L. Zollo, N. Vitiello, N.G. Aracil, S.R. Soekadar, Feasibility and safety of shared EEG/EOG and vision-

guided autonomous whole-arm exoskeleton control to perform activities of daily living. Sci. Rep. **8**, 10823 (2018)

9. J.L. Collinger, B. Wodlinger, J.E. Downey, W. Wang, E.C. Tyler-Kabara, D.J. Weber, A.J. McMorland, M. Velliste, M.L. Boninger, A.B. Schwartz, High-performance neuroprosthetic control by an individual with tetraplegia. Lancet **381**, 557–564 (2013)

10. B.H. Dobkin, Brain-computer interface technology as a tool to augment plasticity and outcomes for neurological rehabilitation. J. Physiol. **579**, 637–642 (2007)

11. A.R. Donati, S. Shokur, E. Morya, D.S. Campos, R.C. Moioli, C.M. Gitti, P.B. Augusto, S. Tripodi, C.G. Pires, G.A. Pereira, F.L. Brasil, S. Gallo, A.A. Lin, A.K. Takigami, M.A. Aratanha, S. Joshi, H. Bleuler, G. Cheng, A. Rudolph, M.A. Nicolelis, Long-term training with a brain-machine interface-based gait protocol induces partial neurological recovery in paraplegic patients. Sci. Rep. **6**, 30383 (2016)

12. V.L. Feigin, M.H. Forouzanfar, R. Krishnamurthi, G.A. Mensah, M. Connor, D.A. Bennett et al., Global and regional burden of stroke in 1990–2010: findings from the global burden of disease study 2010. Lancet **382**, 1–12 (2013)

13. L.R. Hochberg, D. Bacher, B. Jarosiewicz, N.Y. Masse, J.D. Simeral, J. Vogel, S. Haddadin, J. Liu, S.S. Cash, P. van der Smagt, J.P. Donoghue, Reach and grasp by people with tetraplegia using a neurally controlled robotic arm. Nature **485**(7398), 372–375 (2012)

14. M.E. Johanson, J.P. Jaramillo, C.A. Dairaghi, W.M. Murray, V.R. Hentz, Multicenter survey of the effects of rehabilitation practices on pinch force strength after tendon transfer to restore pinch in tetraplegia. Arch. Phys. Med. Rehabil. **97**(6 Suppl), S105–S116 (2016)

15. A. Ramos-Murguialday, D. Broetz, M. Rea, L. Läer, O. Yilmaz, F.L. Brasil, G. Liberati, M.R. Curado, E. Garcia Cossio, A. Vyziotis, W. Cho, M. Agostini, E. Soares, S.R. Soekadar, A. Caria, L.G. Cohen, N. Birbaumer, Brain-machine interface in chronic stroke rehabilitation: a controlled study. Ann. Neurol. **74**, 100–108 (2013)

16. K. Ruddy, J. Balsters, D. Mantini, Q. Liu, P. Kassraian-Fard, N. Enz, E. Mihelj, B. Subhash Chander, S.R. Soekadar, N. Wenderoth, Neural activity related to volitional regulation of cortical excitability. Elife **7**, e40843 (2018)

17. S.R. Soekadar, M. Witkowski, J. Mellinger, A. Ramos Murguialday, N. Birbaumer, L.G. Cohen, ERD-based online brain-machine interfaces (BMI) in the context of neurorehabilitation: optimizing BMI learning and performance. IEEE Trans. Neural Syst. Rehabil. Eng. **19**, 542–549 (2011)

18. S.R. Soekadar, N. Birbaumer, M.W. Slutzky, L.G. Cohen, Brain-machine interfaces in neurorehabilitation of stroke. Neurobiol. Dis. **83**, 172–179 (2015)

19. S.R. Soekadar, M. Witkowski, N. Vitiello, N. Birbaumer, An EEG/EOG-based hybrid brain-neural computer interaction (BNCI) system to control an exoskeleton for the paralyzed hand. Biomed. Tech. **60**, 199–205 (2015)

20. S.R. Soekadar, L.G. Cohen, N. Birbaumer, Clinical brain-machine interfaces, in *Plasticity of cognition in neurologic disorders*, ed. by J. Tracy, B. Hampstead, K. Sathian (Oxford University Press, New York, 2015), pp. 347–362

21. S.R. Soekadar, M. Witkowski, C. Gómez, E. Opisso, J. Medina, M. Cortese, M. Cempini, M.C. Carozza, L.G. Cohen, N. Birbaumer, N. Vitiello, Hybrid EEG/EOG-based brain/neural hand exoskeleton restores fully independent daily living activities after quadriplegia. Sci. Robot. **1**, eaag3296 (2016)

22. S. Toyama, K. Takano, K. Kansaku, A nonadhesive solid-gel electrode for a non-invasive brain–machine interface. Front. Neurol. **3**, 114 (2012)

23. H. Wang, Y. Li, J. Long, T. Yu, Z. Gu, An asynchronous wheelchair control by hybrid EEG–EOG brain–computer interface. Cognit. Neurodyn. **8**, 399–409 (2014)

24. WHO: World health report, Geneva, World Health Organization (2012)

25. M. Witkowski, M. Cortese, M. Cempini, J. Mellinger, N. Vitiello, S.R. Soekadar, Enhancing brain-machine interface (BMI) control of a hand exoskeleton using electrooculography (EOG). J. Neuroeng. Rehabil. **11**, 165 (2014)

26. S.L. Wolf, C.J. Winstein, J.P. Miller et al., The EXCITE investigators. Effect of constraint-induced movement therapy on upper extremity function 3–9 months after stroke: the EXCITE randomized clinical trial. JAMA **296**, 2095–2104 (2006)

Rethinking BCI Paradigm and Machine Learning Algorithm as a Symbiosis: Zero Calibration, Guaranteed Convergence and High Decoding Performance

David Hübner, Pieter-Jan Kindermans, Thibault Verhoeven, Klaus-Robert Müller and Michael Tangermann

Abstract In the past, the decoding quality of brain-computer interface (BCI) systems was often enhanced by independently improving either the machine learning algorithms or the BCI paradigms. We propose to take a novel perspective instead by optimizing the whole system, paradigm and decoder, jointly. To exemplify this holistic idea, we introduce learning from label proportions (LLP) as a new classification approach and prove its value for visual event-related potential (ERP) signals of the EEG. LLP utilizes the existence of subgroups with different label proportions in the data. This leads to a conceptually simple BCI system which combines previously unseen capabilities: (1) it does not require calibration and learns from unlabeled data, (2) under i.i.d. conditions, LLP is guaranteed to obtain the optimal decoder for online data, (3) under violation of stationarity assumptions, LLP can continuously adapt to the changing data, and (4) it can, in practice, replace a traditional supervised decoder when combined with an expectation-maximization algorithm.

Keywords BCI · EEG · ERP · LLP · Learning from label proportions · Unsupervised learning

David Hübner, Pieter-Jan Kindermans—These authors contributed equally.

D. Hübner · M. Tangermann (✉)
Albert-Ludwigs-Universität Freiburg, Freiburg im Breisgau, Germany
e-mail: michael.tangermann@blbt.uni-freiburg.de

P.-J. Kindermans · K.-R. Müller
Berlin Institute of Technology, Berlin, Germany

T. Verhoeven
Ghent University, Ghent, Belgium

K.-R. Müller
Korea University, Seongbuk-gu, Seoul, Korea

Max Planck Institute for Informatics, Saarbrücken, Germany

© The Author(s), under exclusive licence to Springer Nature Switzerland AG 2019   63
C. Guger et al. (eds.), *Brain-Computer Interface Research*,
SpringerBriefs in Electrical and Computer Engineering,
https://doi.org/10.1007/978-3-030-05668-1_6

1   Introduction

One major difficulty in brain-computer interfacing is finding a well-tuned classifier to reliably detect the subject's intention or attentional state. This is a notoriously challenging problem due to the poor signal-to-noise ratio (SNR) in the data, high variance across subjects and non-stationary effects even within single sessions [1, 2]. A variety of approaches have been undertaken to improve classification results.

The first group of approaches aims at improving the SNR of the BCI paradigms. This can be achieved by using new control signals or by improving existing paradigms. The best non-invasive signal quality is generally obtained for visual stimuli such as visual event-related potentials (ERPs). ERPs are time-locked amplitude changes observable for instance in the EEG signal after an internal or external event (such as a visual stimulus) [3]. An example application for ERP-based BCIs is the classical P300-speller by Farwell and Donchin [4]. For visual ERP paradigms, many researchers explored the option of increasing the SNR by improving the salience of the stimuli [5–8] or avoiding confusion between different stimuli [9, 10]. Other visual paradigms rely on an event-related synchronization of the brain to a presented flickering frequency, e.g. with code-modulated potentials (c-VEP) [11] or steady-state visual evoked potentials (SSVEP) [12].

Auditory ERPs were also successfully exploited for controlling BCI paradigms [13, 14], although their signals are generally less informative than visual ones [15]. Besides ERPs, other paradigms rely on oscillatory features of voluntary events (e.g. motor imagery [16]) or on changes in slow cortical potentials [17] which requires intensive user training.

The second group of approaches aims at improving the machine learning model [18–22] to extract more information from the neural data. This includes transfer learning where information is transferred across sessions of the same subject [23] or different subjects [24–27] and unsupervised learning where the decoder is updated with unlabeled data (where the user's intentions are unknown) to adapt an existing classifier [2, 28] or to train a classifier from scratch [29, 30]. Other approaches aim at optimizing the human-computer interaction protocol [31–33] or focus on user skill learning [34].

The previous paragraphs show that a myriad of approaches has been explored in BCI research. In our book chapter, we want to argue that these approaches should not be pursued independently of each other. Instead, optimizing the classification results should be considered as a holistic problem. The BCI field is special from a machine learning point of view, as interventions into the experimental protocol will directly change the structure of the recorded data. Depending on the machine learning model, different data constraints may be required, which—to some extent—can be realized by tuning the paradigm accordingly. In a nutshell, you can create your own machine learning problem in BCI. We want to exemplify this idea with an extremely simple, yet very powerful approach: Depending on the application, classes may occur in different proportions in different subgroups of the data. In an ERP context, we propose to use these label proportions to extract information about non-attended (*non-target*) and attended (*target*) ERP responses. With this information, we can design a powerful unsupervised classifier that:

1. can learn the brain signals from scratch without label information in about 4 min for healthy young subjects in a visual ERP speller,
2. is guaranteed to find a good decoder,
3. can—in practice—compete with a supervised classifier after a short ramp-up phase,
4. can adapt to changing signals over time, and
5. outperforms other unsupervised approaches.

This fills the need for unsupervised adaptation as one of the six '*key challenges for BCI deployment outside the lab*' [35]. It leads towards true plug and play systems and helps to provide stable feedback over time which is important for skill learning [36].

2 Methods

Our new approach involves modifications of the paradigm and of the classifier. We will provide the details thereof in the following two sections.

Paradigm. The traditional row-column visual spelling paradigm by Farwell and Donchin [4] was modified in four ways to realize unsupervised learning with guarantees:

1. The spelling matrix was extended by adding 10 hashtag symbols ('#'). These *visual blank* symbols should be ignored by the user and as such, serve as guaranteed non-targets [37].
2. The highlighting events were subdivided into two subsequences. The first one primarily highlights ordinary characters (except '#' symbols) while the second one mostly highlights visual blank symbols ('#'). This subdivision can be seen as similar to the division in row and column highlighting events [37] and leads to different label proportions for these two subsequences.
3. The highlighting of symbols was performed in pseudo-random groups instead of rows and columns [10].
4. We used a trichromatic rotating grid overlay for highlighting which has shown stronger and more class-discriminative ERP components than traditional brightness highlighting [5].

The second modification is the key change for LLP to work. The first one is added to make LLP easier and more effective. We adopted the latter two as they have shown to increase the SNR, but they are not essential for the new approach. A comparison of the original P300 speller highlighting scheme and that of our paradigm can be seen in Fig. 1.

Classifier. The goal in ERP-based paradigms is to distinguish attended events (targets) from non-attended ones (non-targets). In a previous work, an unsupervised

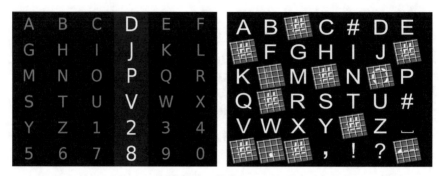

Fig. 1 The original P300 speller (**left**) and our version (**right**) including the visual blank symbols '#', subset highlighting and trichromatic grid overlay

expectation-maximization (EM) algorithm for ERP decoding was introduced by Kindermans et al. and tested in a challenging online auditory BCI setting [29, 30]. We propose to combine the EM algorithm with the newly introduced learning from label proportions (LLP) [38] for decoding ERP signals [37], resulting in the MIX classifier [39]. We summarize the three different classification approaches in Table 1 and explain the details of LLP in the following paragraph.

The LLP classifier utilizes the fact that the two different highlighting sequences have different, but known target to non-target ratios due to a different number of

Table 1 Overview of unsupervised classification methods for ERP decoding

	Expectation—Maximization (EM) [29]	Learning from Label Proportions (LLP) [37]	Mixing (MIX) model estimators [39]
Summary	Probabilistic approach choosing model parameters to maximize data log likelihood	Deterministic approach based on label proportions in the data	Combination of EM and LLP
Conceptual idea	Heavily exploits the structure contained in the data by alternatively estimating a latent variable (= attended symbol) and optimizing the model parameters	Relies on subsets in the data with different class proportions. These subsets can be created for an ERP speller by adding visual blank symbols to the spelling interface (see Fig. 1)	The MIX mean estimation is a convex combination of the other two mean estimations: $\mu_{MIX}(\gamma) = (1 - \gamma)\mu_{LLP} + \gamma\mu_{EM}$ where the mixing coefficient γ is found analytically
Performance	Fast convergence if initialized well and the data has high SNR	Slow, but theoretically guaranteed convergence	Reliable, fast convergence. The dependence upon a good initialization is small

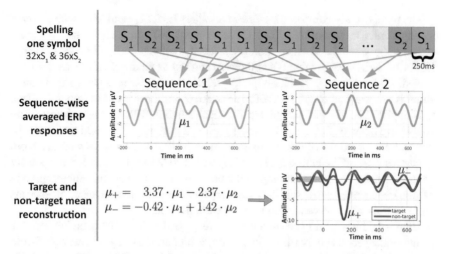

Fig. 2 LLP works by utilizing the existence of two subsequences (S_1, S_2) which have different (and known) target to non-target ratios resulting in different mean responses μ_1 and μ_2 of the two sequences. Knowing these ratios, one can compute the target mean μ_+ and non-target mean μ_- as a linear combination of the sequence means. The coefficients can be derived from the construction of the sequences

visual blanks ("#") and non-blanks (normal characters) being highlighted for each sequence. These ratios are known by construction and can be used to derive a simple linear system of two equations. Solving it delivers an estimate of the target and non-target class means (see Fig. 2). The MIX method then combines these mean estimations with the EM means in a data-driven approach. For classification, we multiply the inverse of the estimated global spatial covariance matrix (estimated on all data disregarding label information) with the target and non-target mean estimates to obtain a linear classifier that is closely related to the LDA classifier [19]. More details are given in [37, 39].

Subjects and Experiments. To evaluate the new unsupervised classification methods, we use data from two EEG studies. In the first study, 13 young healthy subjects copy-spelled a German sentence with 63 symbols three times [37]. Feedback was given by LLP. The data is freely available at the Zenodo Database (http://doi.org/10.5281/zenodo.192684). We used the first 35 letters of each sentence to compare the performance of LLP, EM and MIX and a supervised classifier in a simulation that emulates an online experiment. As the MIX and EM performance depends on the initialization, we randomly initialized each classifier 10 times per sentence. For the deterministic LLP and the supervised classifier, only one classifier was initialized. Matlab code for the unsupervised methods is available in the following Github repository: https://github.com/DavidHuebner/Unsupervised-BCI-Matlab.

The second study compared the three classifiers EM, LLP and MIX with 6 young healthy adults in an online EEG-BCI experiment [40]. In a single session, each subject copy-spelled a German sentence with 35 symbols once with each classifier (LLP, EM, MIX) and received the predicted symbol as feedback after each trial.

Both studies used a visual highlighting scheme with 68 events per symbols and a stimulus onset asynchrony (SOA) of 250 ms. All classifiers were randomly initialized without prior knowledge and updated after each symbol. Although label information would have been available due to copy-spelling, it was never accessed by the classifiers and only used for offline evaluation. In both studies, data were recorded from 31 passive Ag/AgCl electrodes (EasyCap) with 1 kHz sampling rate. Signals were preprocessed by bandpass filtering in the band from 0.5 to 8 Hz and subsampled to 100 Hz. Per channel, extracting the mean amplitudes in 6 intervals located 50–700 ms after stimulus onset served as features for classification.

The visual highlighting scheme led to a very good SNR in the data, but the data quality of the second study (the online study) is higher. It has a grand average 5-fold cross-validated shrinkage-regularized supervised LDA target versus non-target performance of 98.57% against 97.01% in the first study measured by AUC (see below for more details on the metric).

3   Results

To compare the three different unsupervised classifiers, we assessed how well they can discriminate target from non-target events. This can be measured by the area under the curve (AUC) of the receiver-operator curve. The chance level is 50% and perfect accuracy (meaning that the classifier can correctly classify each target or non-target as such) is 100%. Figure 3 shows the results of the simulated online experiment together with the results of the conducted online study. The following observations can be made:

1. LLP initially outperforms EM with a smaller variance across subjects/initializations. EM behaves more dichotomously: depending on the initialization, it can quickly find a very good decoder or fail to do so for a prolonged time period.
2. The MIX method significantly surpasses the other two methods by quickly reaching very high decoding quality. On average, the classification accuracy is above 85% after only 10 symbols (corresponding to around 4 min of unsupervised training time). This accuracy level is already sufficient for reliably controlling the application in most scenarios (see also Fig. 4).
3. Comparing the unsupervised MIX versus a supervised regularized linear discriminant analysis (rLDA) [19], one can observe that the unsupervised MIX method initially gets outperformed, but behaves more similar when more (unlabeled) training data is available. We want to emphasize here that the supervised accuracy estimation is extremely optimistic, as online scenarios require a time

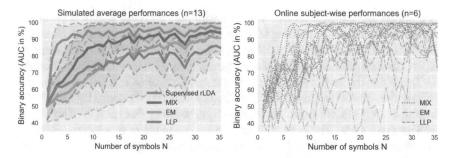

Fig. 3 Comparison of single-trial performances of three unsupervised classifiers and a supervised classifier. In the left subplot, thick lines and the shaded areas bounded by dashed lines depict the grand average binary target versus non-target AUC accuracy over subjects and the mean ±34% areas of the simulated performances, respectively. The right plot shows the subject-wise results of the online study with 6 subjects. All classifiers were trained on the first $N - 1$ symbols and tested on the Nth symbol

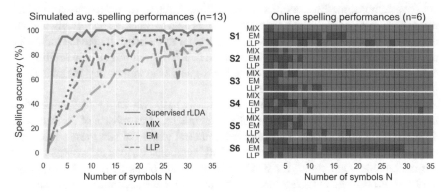

Fig. 4 Comparison of spelling accuracies. The left plot shows the simulated average spelling accuracies. The right plot shows individual spelling outcomes as observed in the online study. Green rectangles denote incorrectly spelled symbols, while gray rectangles denote correctly spelled characters

    point where the supervised calibration phase is ended and from which on no more labels are available, such that no more training can take place. As no such explicit stop-of-training point can be determined in this evaluation, the reported accuracy for the supervised method can be seen as an upper performance bound.

4. The simulation results are congruent with the results obtained in the online study. This can be concluded because almost all traces of the online study are within the confidence band of the simulations. As confidence band, we plotted the mean ±34% of the data which corresponds to a 1σ—band of a normal distribution.

    We also compared the three unsupervised classifiers and the supervised classifier in terms of spelling accuracy, i.e. the percentage of correctly spelled symbols. We show the grand average results of the simulation and results from the online study in Fig. 4. The results underline the claim that the unsupervised MIX method surpasses its unsupervised competitors and reaches a very high decoding ability after only 10 symbols of unsupervised training with an average spelling accuracy of above 80% in the simulation (Fig. 4 left) and 100% in the online experiment (Fig. 4 right). Here, the supervised method shows performance superior to MIX, at least until around symbol 20. This is mostly due to a poor MIX performance on two subjects (which have lower SNR) in the simulation data set. For these two subjects, the supervised method is able to more quickly reach a target versus non-target accuracy that leads to many correct spelling predictions. The very high target versus non-target accuracies, which are also obtained by the MIX method, are not utilized efficiently in this spelling scenario with 68 highlighting events per trial.

4 Discussion and Conclusion

We have demonstrated how a simple modification of a standard BCI paradigm enables a new unsupervised classification method that is able to quickly and reliably learn to decode the user's intended action without any prior knowledge. It uses the unlabeled data almost as efficiently as a supervised method. This algorithm can help to eliminate calibration sessions and adapt to changing signals over time—key ingredients to increase the usability of BCI systems.

It is important to realize that our approach is based on an interaction between the paradigm and the classifier. The LLP principle requires subsets of the data with different label proportions. We presented a natural way to obtain these subsets for a spelling application by adding visual blank symbols. Various extensions thereof are imaginable. One could, for instance, remove the blank symbols in a spelling application once the classifier has ramped up. In other applications, these subsets might already be available. For instance, in a two-step selection protocol, the number of symbols may differ between both steps leading to different target to non-target ratios. One could also get the different ratios from the observation that some letters are more frequently used than others. To enable a symbiosis between the machine learning model and paradigm, one should consider restructuring existing two-step protocols like AMUSE [13] or Hex-o-spell [41] to group infrequent items (letters) together or by changing the number of symbols per selection step. These ratios are more difficult to create in motor imagery scenarios, as fewer events occur over time and the data generally contain less structure.

We see the presented unsupervised approaches as another building block for BCI plug and play systems. Each of the three can be extended by various other techniques—most importantly, the initial ramp-up phase with lower performance can be mitigated using methods from transfer learning where information is transferred across subjects. One can also initially reduce the number of utilized EEG channels as a reduced feature dimensionality has been shown to accelerate the initial learning stage [42]. Other extensions like dynamic stopping (stop the highlighting sequences once the classifier is certain about its decision) [43] or adding a language model for spelling will further increase the usability by improving the information transfer over time as demonstrated by Schreuder et al. [14] and Kindermans et al. [27]. Ultimately, we think that successful BCI systems require a careful analysis of the required neural data structure together with an intelligent fusion of neural and contextual information.

Acknowledgements DH and MT thankfully acknowledge the support by BrainLinks-BrainTools Cluster of Excellence funded by the German Research Foundation (DFG), grant number EXC 1086. DH and MT further acknowledge the bwHPC initiative, grant INST 39/963-1 FUGG. PJK gratefully acknowledges funding from the European Union's Horizon 2020 research and innovation programme under the Marie Sklodowska-Curie grant agreement NO. 657679. TV thankfully acknowledges financial support from the Special Research Fund of Ghent University. KRM thanks DFG (DFG SPP 1527, MU 987/14-1) and the Federal Ministry for Education and Research (BMBF No. 2017-0-00451) as well as support by the Brain Korea 21 Plus Program by the Institute for Information and Communications Technology Promotion (IITP) grant (1IS14013A) funded by the Korean government.

References

1. J.R. Wolpaw, N. Birbaumer, D.J. McFarland, G. Pfurtscheller, T.M. Vaughan, Brain–computer interfaces for communication and control. Clin. Neurophysiol. **113**(6), 767–791 (2002)
2. P. Shenoy, M. Krauledat, B. Blankertz, R.P.N. Rao, K.-R. Müller, Towards adaptive classification for BCI. J. Neural Eng. **3**(1), R13 (2006)
3. S.J. Luck, *An Introduction to the Event-Related Potential Technique* (MIT Press, 2014)
4. L.A. Farwell, E. Donchin, Talking off the top of your head: toward a mental prosthesis utilizing event-related brain potentials. Electroencephalogr. Clin. Neurophysiol. **70**(6), 510–523 (1988)
5. M. Tangermann et al., Optimized stimulation events for a visual ERP BCI. Int. J. Bioelectromagnetism **13**(3), 119–120 (2011)
6. M. Tangermann et al., Data driven neuroergonomic optimization of BCI stimuli, in *Proceedings of 5th International BCI Conference Graz* (2011), pp. 160–163
7. T. Kaufmann, S.M. Schulz, C. Grünzinger, A. Kübler, Flashing characters with famous faces improves ERP-based brain–computer interface performance. J. Neural Eng. **8**(5), 056016 (2011)
8. B. Hong, F. Guo, T. Liu, X. Gao, S. Gao, N200-speller using motion-onset visual response. Clin. Neurophysiol. **120**(9), 1658–1666 (2009)
9. G. Townsend et al., A novel P300-based brain–computer interface stimulus presentation paradigm: moving beyond rows and columns. Clin. Neurophysiol. **121**(7), 1109–1120 (2010)
10. T. Verhoeven, P. Buteneers, J.R. Wiersema, J. Dambre, P.J. Kindermans, Towards a symbiotic brain–computer interface: exploring the application–decoder interaction. J. Neural Eng. **12**(6), 066027 (2015)
11. G. Bin, X. Gao, Y. Wang, Y. Li, B. Hong, S. Gao, A high-speed BCI based on code modulation VEP. J. Neural Eng. **8**(2), 025015 (2011)
12. X. Chen, Y. Wang, M. Nakanishi, X. Gao, T.-P. Jung, S. Gao, High-speed spelling with a noninvasive brain–computer interface. Proc. Natl. Acad. Sci. **112**(44), E6058–E6067 (2015)
13. M. Schreuder, B. Blankertz, M. Tangermann, A new auditory multi-class brain-computer interface paradigm: spatial hearing as an informative cue. PLoS ONE **5**(4), e9813 (2010)
14. M. Schreuder, T. Rost, M. Tangermann, Listen, you are writing! Speeding up online spelling with a dynamic auditory BCI. Front. Neurosci. **5**, 112 (2011)
15. S. Gao, Y. Wang, X. Gao, B. Hong, Visual and auditory brain–computer interfaces. IEEE Trans. Biomed. Eng. **61**(5), 1436–1447 (2014)
16. G. Pfurtscheller, C. Neuper, Motor imagery and direct brain-computer communication. Proc. IEEE **89**(7), 1123–1134 (2001)

17. A. Kübler, N. Neumann, J. Kaiser, B. Kotchoubey, T. Hinterberger, N.P. Birbaumer, Brain-computer communication: self-regulation of slow cortical potentials for verbal communication. Arch. Phys. Med. Rehabil. **82**(11), 1533–1539 (2001)
18. K.-R. Müller, M. Tangermann, G. Dornhege, M. Krauledat, G. Curio, B. Blankertz, Machine learning for real-time single-trial EEG-analysis: from brain–computer interfacing to mental state monitoring. J. Neurosci. Methods **167**(1), 82–90 (2008)
19. B. Blankertz, S. Lemm, M. Treder, S. Haufe, K.-R. Müller, Single-trial analysis and classification of ERP components, a tutorial. Neuroimage **56**(2), 814–825 (2011)
20. F. Lotte, M. Congedo, A. Lécuyer, F. Lamarche, B. Arnaldi, A review of classification algorithms for EEG-based brain–computer interfaces. J. Neural Eng. **4**(2), R1 (2007)
21. B. Blankertz, G. Curio, K.-R. Müller, Classifying single trial EEG: towards brain computer interfacing, in *Advances in Neural Information Processing Systems*, ed. by T.G. Dietterich, S. Becker, Z. Ghahramani, vol. 14 (MIT Press, 2002), pp. 157–164
22. B. Blankertz, R. Tomioka, S. Lemm, M. Kawanabe, K.-R. Müller, Optimizing spatial filters for robust EEG single-trial analysis. IEEE Signal Process. Mag. **25**(1), 41–56 (2008)
23. M. Krauledat, M. Tangermann, B. Blankertz, K.-R. Müller, Towards zero training for brain-computer interfacing. PLoS ONE **3**(8), e2967 (2008)
24. S. Fazli, F. Popescu, M. Danóczy, B. Blankertz, K.-R. Müller, C. Grozea, Subject-independent mental state classification in single trials. Neural Netw. **22**(9), 1305–1312 (2009)
25. S. Fazli, S. Dähne, W. Samek, F. Bießmann, K.-R. Müller, Learning from more than one data source: data fusion techniques for sensorimotor rhythm-based brain-computer interfaces. Proc. IEEE **103**(6), 891–906 (2015)
26. V. Jayaram, M. Alamgir, Y. Altun, B. Schölkopf, M. Grosse-Wentrup, Transfer learning in brain-computer interfaces. IEEE Comput. Intell. Mag. **11**(1), 20–31 (2016)
27. P.-J. Kindermans, M. Tangermann, K.-R. Müller, B. Schrauwen, Integrating dynamic stopping, transfer learning and language models in an adaptive zero-training ERP speller. J. Neural Eng. **11**(3), 035005 (2014)
28. C. Vidaurre, M. Kawanabe, P. von Bünau, B. Blankertz, K.-R. Müller, Toward unsupervised adaptation of LDA for brain–computer interfaces. IEEE Trans. Biomed. Eng. **58**(3), 587–597 (2011)
29. P.-J. Kindermans, D. Verstraeten, B. Schrauwen, A bayesian model for exploiting application constraints to enable unsupervised training of a P300-based BCI. PLoS ONE **7**(4), e33758 (2012)
30. P.-J. Kindermans, M. Schreuder, B. Schrauwen, K.-R. Müller, M. Tangermann, True zero-training brain-computer interfacing—an online study. PLoS ONE **9**(7), e102504 (2014)
31. R. Chavarriaga, P.W. Ferrez, J. del R. Millán, To err is human: learning from error potentials in brain-computer interfaces, in *Proceedings of the International Conference on Cognitive Neurodynamics (ICCN)* (2007), pp. 777–782
32. I. Iturrate, R. Chavarriaga, L. Montesano, J. Minguez, J. del R. Millán, Teaching brain-machine interfaces as an alternative paradigm to neuroprosthetics control. Sci. Rep. **5**(13893), 1–10 (2015)
33. T.O. Zander, L.R. Krol, N.P. Birbaumer, K. Gramann, Neuroadaptive technology enables implicit cursor control based on medial prefrontal cortex activity. Proc. Natl. Acad. Sci. **113**(52), 14898–14903 (2016)
34. F. Lotte, C. Jeunet, Towards improved BCI based on human learning principles, in *The 3rd International Winter Conference on Brain-Computer Interface* (2015), pp. 1–4
35. L.F. Nicolas-Alonso, J. Gomez-Gil, Brain computer interfaces, a review. Sensors **12**(2), 1211–1279 (2012)
36. J.d.R. Millán et al., Combining brain–computer interfaces and assistive technologies: State-of-the-art and challenges. Front. Neurosci. **4**(161), 1–15 (2010)
37. D. Hübner, T. Verhoeven, K. Schmid, K.-R. Müller, M. Tangermann, P.-J. Kindermans, Learning from label proportions in brain-computer interfaces: online unsupervised learning with guarantees. PLoS ONE **12**(4), e0175856 (2017)

38. N. Quadrianto, A.J. Smola, T.S. Caetano, Q.V. Le, Estimating labels from label proportions. J. Mach. Learn. Res. **10**, 2349–2374 (2009)
39. T. Verhoeven, D. Hübner, M. Tangermann, K.-R. Müller, J. Dambre, P.-J. Kindermans, Improving zero-training brain-computer interfaces by mixing model estimators. J. Neural Eng. **14**(3), 036021 (2017)
40. D. Hübner, T. Verhoeven, P.-J. Kindermans, M. Tangermann, Mixing two unsupervised estimators for event-related potential decoding: an online evaluation, in *Proceedings of the 7th International Brain-Computer Interface Meeting 2017: From Vision to Reality* (2017), pp. 198–203
41. K.-R. Müller, B. Blankertz, Toward noninvasive brain-computer interfaces. IEEE Signal Process. Mag. **23**(5), 128–126 (2006)
42. D. Hübner, P.-J. Kindermans, T. Verhoeven, M. Tangermann, Improving learning from label proportions by reducing the feature dimensionality, in *Proceedings of the 7th International Brain-Computer Interface Meeting 2017: From Vision to Reality* (2017), pp. 186–191
43. M. Schreuder, J. Höhne, B. Blankertz, S. Haufe, T. Dickhaus, M. Tangermann, Optimizing event-related potential based brain–computer interfaces: a systematic evaluation of dynamic stopping methods. J. Neural Eng. **10**(3), 036025 (2013)

Targeted Up-Conditioning of Contralesional Corticospinal Pathways Promotes Motor Recovery in Poststroke Patients with Severe Chronic Hemiplegia

K. Takasaki, F. Liu, M. Ogura, K. Okuyama, M. Kawakami, K. Mizuno,
S. Kasuga, T. Noda, J. Morimoto, M. Liu and J. Ushiba

Abstract Impairment of shoulder elevation in poststroke hemiplegia is a debilitating condition with no evidence-based, accessible treatment. This study evaluated the safety and efficacy of direct brain control of advanced exoskeleton robotics as a physiotherapeutic intervention. Poststroke patients with severe chronic hemiplegia participated in a physiotherapeutic intervention with movement support aided by online decoding of contralesional primary motor cortex activity and exoskeleton shoulder robotics. Participants engaged in 1 h of daily exercise for 7 consecutive days, which promoted lateralized motor-related electroencephalogram (EEG) responses to the contralesional side and the appearance of a transcranial magnetic stimulation-evoked potential in the paralyzed shoulder muscle. Participants gained active range-of-motion in the affected arm, with a flexion movement beyond the standardized minimal clinically important difference. These results suggest that an EEG-based brain-machine interface could facilitate targeted up-conditioning of contralesional corticospinal pathways, resulting in the clinically relevant functional recovery of movement.

Keywords Stroke · Chronic hemiplegia · EEG · TMS · BMI · BCI

K. Takasaki
Graduate School of Science and Technology, Keio University, Kanagawa, Japan

F. Liu · M. Ogura · K. Okuyama · M. Kawakami · K. Mizuno · M. Liu
Department of Rehabilitation Medicine, Keio University, Tokyo, Japan

S. Kasuga · J. Ushiba (✉)
Faculty of Science and Technology, Keio Institute of Pure and Applied Science (KiPAS), Keio University, Kanagawa, Japan
e-mail: ushiba@brain.bio.keio.ac.jp

T. Noda · J. Morimoto
Department of Brain Robot Interface, Advanced Telecommunications Research Institute International, Kyoto, Japan

1   Introduction

Following a stroke injury, the human brain initiates a dynamic process of functional remodeling in neural circuitry. However, this process typically halts when it reaches a chronic stage, approximately 6 months poststroke. The efficacy of rehabilitation in this chronic stage of stroke remains debatable. Although many approaches have been used to improve recovery of upper extremity (UE) function, few interventions are available for patients with severe hemiparesis in the chronic phase [1].

Recent studies have demonstrated that motor exercise supported by an electroencephalogram (EEG)-based brain-machine interface (BMI) facilitates the resumption and shaping of the neural remodeling process [2, 3]. Such remodeling processes occur even in the non-plastic chronic stage, continuing to unmask the remaining circuitry of the original corticospinal pathways in the ipsilesional hemisphere [2, 3]. Randomized controlled trials have suggested that the promotion of functional recovery of finger and arm movements is clinically useful [4, 5].

However, the remodeling process of the ipsilesional primary motor cortex (M1) for finger motor recovery has been shown to impede shoulder movement recovery, as enlargement of the motor areas associated with finger control can lead to erosion of motor areas responsible for shoulder movement [6]. To avoid this competitive reinnervation process in the ipsilesional M1, we targeted the corticospinal tract from the contralesional hemisphere (i.e., ipsilateral to the paralyzed arm) for shoulder motor recovery, since shoulder muscles are bihemispherically innervated [7, 8]. In this study, we monitored the participants' contralesional EEG, and event-related desynchronization (ERD) of the sensorimotor rhythm (SMR) by means of kinesthetic motor imagery of shoulder elevation was used to trigger actions of the robotic exoskeleton and neuromuscular electrical stimulation. Since ERD observed from the hemisphere ipsilateral to the imaged arm represents corticospinal tract excitability on the same side [9], we designed this intervention with the aim of achieving targeted up-conditioning of contralesional corticospinal pathways for shoulder flexion movement.

2   Methods

2.1   Study Design and Protocol

This trial was designed as a proof-of-concept study testing the safety and preliminary efficacy of the approach (phase 2) [10]. Eight poststroke patients with severe hemiplegia in the chronic stage were recruited to this study (4 males/4 females; age 58.4 ± 11.1 years; time from stroke onset 29 ± 25.9 months; Fugl-Meyer assessment [FMA] score 16.8 ± 4.4). The study was approved by the institutional ethics review board (approval 20140442), registered with the University Hospital Medical Information Network (UMIN) (UMIN Clinical Trials Registry: No. UMIN000017525), and was

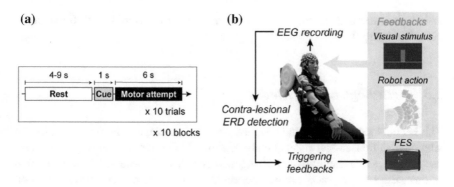

Fig. 1 Shoulder elevation exercise with direct brain-controlled exoskeleton robotics. **a** Time scheme of the training protocol. **b** Outline of the developed brain-machine interface system. ERD: event-related desynchronization; FES: functional electrical stimulation

performed in accordance with the Declaration of Helsinki after written, informed consent was obtained from the participants.

Figure 1 summarizes the setup of the BMI intervention. The interventions were carried out for 7 consecutive days without conventional therapy, and clinical and neurophysiological measurements were performed 1 day before and after the intervention. In the intervention, participants were asked to attempt shoulder flexion for 6 s after a short notice from a cue (1 s). Four to nine seconds of rest was allowed afterward. The time sequence was alerted by visual cues on a computer screen. This motor task consisted of 10 runs of 10 trials, for a total of 100 trials per day (Fig. 1a). During the motor attempt, SMR-ERD over the contralesional hemisphere was obtained, and the participants were given visual bar feedback of its amplitude. When the participants maintained ERD over 30% for 1 s during the motor attempt period, the robotic assistance of shoulder flexion and neuromuscular electrical stimulation (100 Hz, 1 ms), with an intensity set to the motor threshold of the anterior-deltoid, was applied as feedback (Fig. 1b).

2.2 Functional Assessments

One medical doctor assessed all patients. The arm (UE) section of the FMA (FMA-UE) [11, 12] was used for the primary outcome measure. In the present analysis, the Minimal Clinically Important Difference (MCID) value was determined as 5 from the initial FMA score [13, 14]. Other functional outcome measures included the Stroke Impairment Assessment Set (SIAS) [15]. The SIAS is a standardized measure of stroke impairment, consisting of 22 subcategories, and has excellent interrater reliability [16, 17]. Motor functions of the paretic UE are tested with the knee-mouth test and the finger test. They are rated from 0 to 5, with 0 indicating complete paralysis and 5 no paresis.

2.3 Neurological Assessments

The lateralization of ERD and the motor-evoked potential (MEP) induced by the single-pulse transcranial magnetic stimulation (TMS) of the anterior-deltoid were used to assess the neurological changes in M1. To calculate ERD as a function of time, EEG data of each trial was collected in the following manner: (1) common averaged filter, bandpass (0.01–70 Hz) and notch (50 Hz) filtering with a Butterworth filter of order 4, (2) segmentation for 1 s of EEG data and 90% overlap time window, (3) Fast Fourier transform (FFT) with Hann window function, (4) power spectrum density calculation, (5) ERD transformation using the following equation:

$$\text{ERD}(f, t) = \frac{A(f, t) - R(f)}{R(f)} \times 100,$$

where variable A was the power spectrum of the recorded EEG at frequency f at time t with reference to the onset of motor imagery, and R is the power spectrum of 2-s between 1 s after onset of rest to 1 s before the end of the rest period. By using this equation, ERD was calculated as a negative number and event-related synchronization (ERS) was calculated as a positive value.

A 128-channel whole-head EEG (Net Station, Electrical Geodesics, Inc., Eugene, OR, USA) was used to assess hemispheric lateralization of ERD. The Laterality Index value (LI) was computed using the following formula,

$$\text{Laterality Index (LI)} = \left(\frac{cERD - iERD}{|cERD| + |iERD|} \right),$$

where cERD and iERD are the values of the contra- and ipsilesional ERD, respectively. For the patients with right-sided hemiplegia, iERD was calculated using the mean ERD amplitude across the C4 channel and 6 channels around C4. cERD was calculated using the mean ERD amplitude across the C3 channel and 6 channels around C3. This correspondence was reversed to calculate iERD and cERD in case of left-sided hemiplegia.

In 6 of 8 participants who did not have contraindications for TMS, MEPs of the paralyzed anterior deltoid were measured during shoulder flexion at 20% of maximum isometric voluntary contraction. TMS was applied over the contralesional M1 at 110% active motor threshold. The EMG activity from the paralyzed anterior deltoid muscle was recorded with a pair of Ag/AgCl surface electrodes. The amplified and bandpass-filtered (0.1–2 kHz) raw EMG signal was digitized at a 20 kHz sampling rate and stored for offline analysis.

3 Results

From the functional assessments, FMA-UE was significantly improved after the intervention (paired t-test, $p < 0.01$). All patients improved FMA and 6 of 8 patients exceeded the MCID values (Fig. 2). The mean change in the FMA score after the intervention was 6.63 ± 3.16.

The amplitude of contralesional SMR-ERD and LI of SMR-ERD is shown in Fig. 3. The amplitude of SMR-ERD in M1 of the contralesional hemisphere (Fig. 3a) was significantly increased after the interventions (Wilcoxon signed-rank test, $p < 0.01$). The group analysis of hemispheric lateralization of ERD using the LI value (Fig. 3b) showed that the contralesional ERD became dominant after BMI interventions (Wilcoxon signed-rank test, $p < 0.01$). MEPs became more apparent after the interventions in 3 of the 6 tested participants (Patients 4, 6, 8 in Table 1). MEPs did not change in Patient 1 and MEPs did not appear in Patients 2 and 3.

Fig. 2 FMA gain in eight patients with stroke

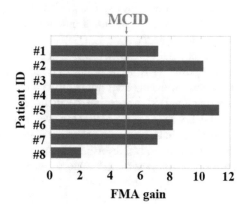

Fig. 3 Amplitude of contralesional event-related desynchronization (ERD)

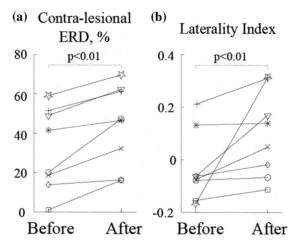

Table 1 Change of MEPs

Patient ID	ipsi-lateral MEP (mV)	
	Pre	Post
1	0.18	0.18
4	0.27	0.32
6	0.65	2.14
8	N/A	0.24

MEP motor-evoked potential

4 Discussion

We developed a novel intervention method for achieving shoulder motor exercise in stroke patients who have severe paralysis, with the aim of achieving targeted up-conditioning of contralesional corticospinal pathways related to shoulder flexion movement. In this pilot study, we found a significant improvement in the FMA-UE score, as a primary functional outcome measure, in poststroke patients with severe chronic hemiplegia.

The clinical efficacy of robotic rehabilitation for stroke patients has previously been reported [14]. A randomized controlled trial (RCT) of BMI has shown FMA score improvements after intervention with a robotic exoskeleton for the recovery in arm function [18]. According to a meta-analysis, the pooled average improvement in FMA-UE was 3.7 ± 0.5 [19]. However, in this study, we confirmed an effect size of almost 1.9 times by our intervention approach, although the sample size was small in the present study. Nevertheless, our participants demonstrated significant motor recovery beyond the MCID value, suggesting the strong potential of this intervention as an engaging treatment with no currently identified side-effects.

The present study shows that SMR-ERD was lateralized to the contralesional side of the hemisphere. The ipsilesional MEP, as assessed by TMS, became more apparent in 3 of 4 tested participants. These signs suggested the up-conditioned ipsilateral cortico-spinal excitability after the BMI intervention. SMR-ERD has a known EEG signature representing the excitability of the sensorimotor cortex. Furthermore, MEP reflects M1 excitability. Several groups have reported the positive correlation between MEP and ERD in the contralateral hemisphere for unilateral finger motor imagery [20]. More recently, Hasegawa et al. reported that MEP recorded from proximal muscles also positively correlated with the ERD in the ipsilateral hemisphere [9]. This evidence further indicates that our data represent the up-conditioning of the contralesional corticospinal tract through our intervention.

In conclusion, in the current study, we have demonstrated the feasibility and the clinical safety of a novel BMI intervention. Moreover, the clinical effects exceeded MCID values in the majority of patients. In future, RCTs of a large number of participants are required to evaluate the clinical efficacy of targeted up-conditioning of contralesional corticospinal pathways by this BMI intervention in phase 3 clinical trials.

Acknowledgements An original study on shoulder movement restoration by using BMI in post-stroke hemiplegia was submitted to the Annual BCI Award 2017 hosted by g.tec., Austria, and was nominated among the top 12 research projects. This article is a reprinted version of the nominated article, with some supplementary explanation. A short summary of this study was previously introduced in IEEE eNewsLetter in 2018.

This study was partially supported by the Strategic Research Program for Brain Sciences from the Ministry of Education, Culture, Sports, Science, and Technology of Japan.

References

1. L. Oujamaa, I. Relave, J. Froger et al., Rehabilitation of arm function after stroke. Literature review. Ann. Phys. Rehabil. Med. **52**, 269–93 (2009)
2. K. Shindo, K. Kawashima, J. Ushiba et al., Effects of neurofeedback training with an electroencephalogram-based brain-computer interface for hand paralysis in patients with chronic stroke: a preliminary case series study. J. Rehabil. Med. **43**, 951–57 (2011)
3. J. Ushiba, S. Soekadar, Brain-machine interfaces for rehabilitation of poststroke hemiplegia. Prog. Brain Res. **228**, 163–83 (2016)
4. A. Ramos-Murguialday, D. Broetz, M. Rea et al., Brain-machine-interface in chronic stroke rehabilitation: a controlled study. Ann. Neurol. **74**, 100–8 (2013)
5. F. Pichiorri, G. Morone, M. Petti et al., Brain-computer interface boosts motor imagery practice during stroke recovery. Ann. Neurol. **77**, 851–65 (2015)
6. R. Chen, L.G. Cohen, M. Hallett, Role of the ipsilateral motor cortex in voluntary movement. Can. J. Neurol. Sci. **24**, 284–91 (1997)
7. E.S. Rosenzweig, J.H. Brock, M.D. Culbertson et al., Extensive spinal decussation and bilateral termination of cervical corticospinal projections in rhesus monkeys. J. Comp. Neurol. **513**, 151–63 (2009)
8. P. Bawa, J.D. Hamm, P. Dhillon et al., Bilateral responses of upper limb muscles to transcranial magnetic stimulation in human subjects. Exp. Brain Res. **158**, 385–90 (2004)
9. K. Hasegawa, S. Kasuga, K. Takasaki et al., Ipsilateral EEG mu rhythm reflects the excitability of uncrossed pathways projecting to shoulder muscles. J. Neuroeng. Rehabil. **14**, 85 (2017)
10. B.H. Dobkin, Progressive staging of pilot studies to improve phase III trials for motor interventions. Neurorehabil. Neural. Repair. **23**, 197–206 (2009)
11. A.R. Fugl-Meyer, L. Jääskö, I. Leyman et al., The post-stroke hemiplegic patient. 1. A method for evaluation of physical performance. Scand. J. Rehabil. Med. **7**, 13–31 (1975)
12. D.J. Gladstone, C.J. Danells, S.E. Black, The Fugl-Meyer assessment of motor recovery after stroke: a critical review of its measurement properties. Neurorehabil. Neural. Repair. **16**, 232–40 (2002)
13. S.J. Page, G.D. Fulk, P. Boyne, Clinically important differences for the upper-extremity Fugl-Meyer Scale in people with minimal to moderate impairment due to chronic stroke. Phys. Ther. **92**, 791–8 (2012)
14. P. Langhorne, F. Coupar, A. Pollock, Motor recovery after stroke: a systematic review. Lancet Neurol. **8**, 741–54 (2009)
15. N. Chino, S. Sonoda, K. Domen et al., Stroke Impairment Assessment Set (SIAS)—a new evaluation instrument for stroke patients. Jpn. J. Rehabil. Med. **31**, 119–25 (1994)
16. T. Tsuji, M. Liu, S. Sonoda et al., The stroke impairment assessment set: its internal consistency and predictive validity. Arch. Phys. Med. Rehabil. **81**, 863–68 (2000)
17. M. Liu, N. Chino, T. Tuji et al., Psychometric properties of the Stroke Impairment Assessment Set (SIAS). Neurorehabil. Neural. Repair **16**, 339–51 (2002)
18. K.K. Ang, K.S.G. Chua, K.S. Phua et al., A randomized controlled trial of EEG-based motor imagery brain–computer interface robotic rehabilitation for stroke. Clin. EEG Neurosci. **46**, 310–20 (2015)

19. G.B. Prange, M.J. Jannink, C.G. Groothuis-Oudshoorn et al., Systematic review of the effect of robot-aided therapy on recovery of the hemiparetic arm after stroke. J. Rehabil. Res. Dev. **43**, 171–84 (2006)
20. M. Takemi, Y. Masakado, M. Liu et al., Event-related desynchronization reflects downregulation of intracortical inhibition in human primary motor cortex. J. Neurophysiol. **110**, 1158–66 (2013)

Individual Word Classification During Imagined Speech Using Intracranial Recordings

Stephanie Martin, Iñaki Iturrate, Peter Brunner, José del R. Millán, Gerwin Schalk, Robert T. Knight and Brian N. Pasley

Abstract In this study, we evaluated the ability to identify individual words in a binary word classification task during imagined speech, using high frequency activity (HFA; 70–150 Hz) features in the time domain. For this, we used an imagined word repetition task cued with a word perception stimulus, and followed by an overt word repetition, and compared the results across the three conditions. We used support-vector machines, and introduced a non-linear time-realignment in the classification framework—in order to deal with speech temporal irregularities. As expected, high classification accuracy was obtained in the listening (mean = 89%) and overt speech conditions (mean = 86%), where speech stimuli were directly observed. In the imagined speech condition, where speech is generated internally by the patient, results show for the first time that individual words in single trials were classified with statistically significant accuracy. Classification accuracy reached 88% in a two-class classification framework, and average classification accuracy across fifteen word-pairs was significant across five subjects (mean = 58%). The majority of electrodes carrying discriminative information were located in the superior temporal gyrus, inferior frontal gyrus and sensorimotor cortex, regions commonly associated with speech processing. These data represent a proof of concept study for basic decoding of speech imagery, and delineate a number of key challenges to usage of speech imagery neural representations for clinical applications.

S. Martin (✉) · I. Iturrate · J. del R. Millán
Defitech Chair in Brain-Machine Interface, Center for Neuroprosthetics, Ecole Polytechnique Fédérale de Lausanne, Chemin des Mines 9, 1202 Geneve, Switzerland
e-mail: stephanie.martin@epfl.ch

S. Martin · R. T. Knight · B. N. Pasley
Helen Wills Neuroscience Institute, University of California, Berkeley, CA, USA

P. Brunner · G. Schalk
National Center for Adaptive Neurotechnologies, Wadsworth Center, New York State Department of Health, Albany, NY, USA

Department of Neurology, Albany Medical College, Albany, NY, USA

R. T. Knight
Department of Psychology, University of California, Berkeley, CA, USA

© The Author(s), under exclusive licence to Springer Nature Switzerland AG 2019
C. Guger et al. (eds.), *Brain-Computer Interface Research*,
SpringerBriefs in Electrical and Computer Engineering,
https://doi.org/10.1007/978-3-030-05668-1_8

Keywords Imagined speech · Overt speech · High frequency activity · HFA ·
Electrocorticography

1 Introduction

People with speech production impairments would benefit from a system that can
infer intended speech directly from brain signals. Here, we used direct cortical recording to examine if individual words could be selected during imagined speech within
a binary classification framework. However, despite intense investigation, the neural mechanisms underlying imagined speech remain poorly defined, in part due to
the lack of clear timing of inner speech, the subjective nature and inter-individual
differences in how subject imagine speech.

Functional magnetic resonance imaging studies have shown that imagined speech
activates Wernicke's area [1, 2, 4, 13, 15, 22, 26] and Broca's area [8, 9]—two essential language areas involved in speech comprehension and production, respectively
(see Refs. [17, 18] for reviews). Although traditional brain imaging techniques have
identified anatomical regions associated with imagined speech, these methods lack
the temporal resolution to investigate the rapid temporal neural dynamics during
imagined speech [24]. In contrast, electrocorticography is a direct neural recording
method that allows monitoring brain activity with high spatial, temporal, and spectral
resolution [20].

In this study, we took advantage of the high resolution offered by ECoG to classify
individual during imagined speech, using HFA features in the time domain. However, speech production (both overt and imagined) is subject to temporal variations
(speech onset delays and local stretching/compression) across repetitions of the same
utterance (Fig. 1a, top panel; [19, 25]). As a result, a classifier that assumes fixed
time features may not recognize two trials as belonging to the same class if the neural
patterns were not temporally aligned. To overcome that limitation, we proposed a
new classification framework that accounted for temporal variations during speech
production (overt and imagined) by introducing time realignment in the feature map
generation (Fig. 1a, bottom panel). In particular, we used support-vector machines
[6] to classify individual words in a word pair, and introduced a non-linear time alignment into the kernel to deal with internal speech production variability. We used an
imagined word repetition task cued with a word perception stimulus, and followed
by an overt word repetition (Fig. 1b), and compared the results across the three conditions (listening, overt and imagined speech) in five epileptic patients undergoing
neurosurgery for brain ablation (Fig. 1c).

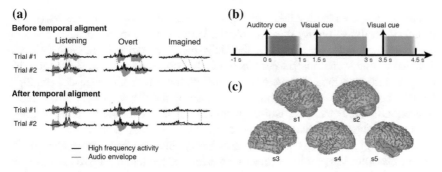

Fig. 1 Single trial neural activity a Examples of two repetitions of the same word for an active electrode in all three different conditions (listening, overt and imagined speech) before (top panel) and after (bottom panel) temporal realignment with dynamic time warping. Black indicates the high frequency activity; red indicates the audio speech envelope. **b** Subjects were presented with an auditory stimulus that indicated one of six individual words. Then, a cue appeared on the screen, and subjects had to imagine hearing the word they had just heard. Finally, a second cue appeared, and subjects had to say the word out loud. Shaded areas represent the intervals extracted for classification. **c** Grid locations for each subject were overlaid on cortical surface reconstructions of each subject's MRI scan

2 Results

We used support-vector machines [6] to perform pair-wise classification of different individual words in the three different speech conditions (see Ref. [12] for details). We first extracted the high frequency activity using bandpass filtering in the 70–150 Hz range, then extracted the envelope using the Hilbert transform. We then extracted epochs from 100 ms before speech onset to 100 ms after speech offset for both listening and overt speech conditions. Average word length for both listening and overt speech conditions were 800 ms $\pm$ 20 and 766 ms $\pm$ 84, respectively. In the imagined speech condition, due to the lack of speech output, we extracted 1500 ms epochs starting at cue onset. In the listening condition, the auditory stimuli were time-locked across repetitions (Fig. 1a). Alternatively, in the overt speech condition, temporal irregularities in the speech onset and word duration were observed across repetitions of the same utterance (Fig. 1a). Because high frequency neural activity is known to track the speech envelope [10, 11, 14, 16], we assumed that temporal variations in overt and imagined speech would also be represented by the measured neural responses.

In our study, to deal with speech temporal irregularities, we incorporated time alignment in the kernel computation. The Gaussian kernel is a widely used function used in SVM-classification. In this approach, the output of the classifier is based on a weighted linear combination of similarity measures (i.e., Euclidean distance) computed between a data point and each of the support vectors [6]. Here, we used the realigned Euclidean distance, using dynamic time warping (DTW; Refs. [19, 21])—to locally expand or compress two time series, and find their optimal alignment in time.

Then, we computed the Euclidian distance between the realigned time series (DTW-distance), as the similarity measure for the kernel computation. As such, for each electrode separately, we computed the DTW-distance between each pair of trials. This led to one kernel matrix per electrode. To build the final kernel function, we computed the weighted average of the kernel matrices over all electrodes (multiple kernel learning; [5]). The weighting was based on the discriminative power index of each individual electrode, which quantified the difference between the "within" class versus "between" class distances distribution (see Ref. [12]).

In the imagined speech condition, pairwise classification accuracy reached 88.3% for one classification pair in a subject with extensive left temporal coverage (subject 4; Fig. 2a). Eight out of fifteen word-pairs were classified significantly higher than chance level ($p < 0.05$; randomization test; FDR correction), exceeding the number of pairs expected by chance ($0.05 \times 15 = 0.75$). As expected, higher classification accuracy was obtained in the listening and overt speech conditions where speech stimuli were directly observed. For both conditions, pairwise classification accuracy approached 100% in some comparisons, and twelve and fifteen out of fifteen pairs were significantly above chance, respectively ($p < 0.05$; randomization test; FDR correction).

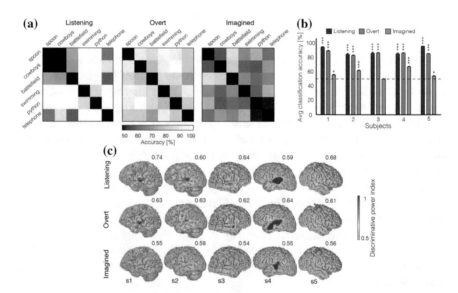

Fig. 2 Classification accuracy. a Pairwise classification accuracy in the testing set for the listening (left panel), overt speech (middle panel) and imagined speech conditions (right panel) for a subject with good temporal coverage (S4). **b** Average classification accuracy across all pairs of words for each subject and condition (listening, overt and imagined speech). Error bars denote SEM. ***$p < 0.001$, **$p < 0.01$, *$p < 0.5$. **c** Discriminative power measured as the areas under the ROC curve (thresholded at $p < 0.05$; uncorrected), and plotted on each individual's brain. Each is scaled to the maximum absolute value of discriminative power index (indicated by the number above each cortical map)

Classification accuracy varied across subjects and pairs of words. In 4 out of 5 subjects, classification accuracy over all word pairs was significant in the imagined speech condition (Fig. 2b; $p < 0.05$; one-sample t-test; FDR correction), while the last subject was not significantly better than chance level (mean $= 49.8\%$; $p > 0.5$; one-sample t-test; FDR correction). For listening and overt speech conditions, classification accuracy over all word pairs was again significant in all four subjects, and ranged between 83.0 and 96.0% ($p < 10^{-4}$; one-sample t-test; FDR correction).

At the population level, average classification accuracy across all pairs was above chance level in all three conditions (Fig. 2b; listening: mean $= 89.4\% \ p < 10^{-4}$; overt speech: mean $= 86.2\%$, $p < 10^{-5}$; imagined speech: mean $= 57.7\%$; $p < 0.05$; one-sample t-tests; FDR correction). A repeated measure 1-way ANOVA with experimental condition as a factor confirmed a difference among conditions ($F_{(2,12)} = 56.3, p < 10^{-5}$). Post-hoc t-tests showed that the mean classification accuracy for listening was not significantly different from the overt speech ($p > 0.1$; two-sample t-test; FDR correction). Both were significantly higher than the imagined speech classification accuracy ($p < 0.005$; two-sample t-test; FDR correction). Although the classification accuracy for imagined speech was lower than for listening and overt speech, the imagery classification results provide evidence that the HFA time course during imagined speech contained information to distinguish pairs of words.

To assess how the brain areas important for word classification vary across experimental conditions, we analyzed the anatomical distribution of the electrodes carrying discriminative information in the three different conditions. For each electrode and condition, we computed a discriminative power index that reflected the predictive power of each electrode in the classification process.

Figure 2c shows the anatomical distribution of the discriminative power index across each condition (heat map thresholded at $p < 0.05$; uncorrected). Overall, the most discriminative information was located in the temporal gyrus, inferior frontal gyrus and sensorimotor gyrus—regions commonly associated with speech processing. Anatomical differences between conditions were assessed for significant electrodes (188 electrodes significant in at least one condition; $p < 0.05$; FDR correction), using an unbalanced Two-Way ANOVA with interactions, with experimental condition (listening, overt and imagined speech) and anatomical region [superior temporal gyrus (STG), inferior frontal gyrus (IFG) and sensorimotor cortex (SMC)] as factors. The main effect of experimental condition was significant [$F_{(2,555)} = 29.1$, $p < 10^{-15}$], indicating that the discriminative information in the classification process was different across conditions. Post-hoc t-tests with Bonferroni correction showed that the overall discriminative power was higher in the listening (mean $= 0.56$) and overt speech condition (mean $= 0.56$) than in the imagined speech (mean $= 0.53$; $p < 10^{-10}$; unpaired two-sample t-test; Bonferroni correction), at the level of single electrodes. The main effect of anatomical region was also significant [$F_{(2,555)} = 7.18$, $p < 0.001$]. Post-hoc t-tests indicated stronger discriminative information in the STG (mean $= 0.55$) than in the inferior frontal gyrus (mean $= 0.54$; $p < 0.05$; unpaired two-sample t-test; Bonferroni correction), but not than the SMC (mean $= 0.54$; $p > 0.05$; unpaired two-sample t-test; Bonferroni correction). The interaction between gyrus and experimental condition was also significant [$F_{(4,555)} = 6.7$; $p < 10^{-4}$].

Specifically, the discriminative power in the STG was higher for listening (mean = 0.57) and overt speech (mean = 0.56) than for imagined speech (mean = 0.53; $p <$ 10^{-10}; unpaired two-sample t-test; Bonferroni correction). In addition, the discriminative power in the sensorimotor cortex was higher in the overt condition (mean = 0.57), than in the listening (mean = 0.54) and imagined conditions (mean = 0.53; $p <$ 0.001; unpaired two-sample t-test; Bonferroni correction). Similarly, the frontal electrodes provided more discriminative information in the overt speech (mean = 0.55) than in the imagined speech condition (mean = 0.53; $p < 10^{-4}$; unpaired two-sample t-test; Bonferroni correction). Post-hoc t-tests also showed that the discriminative power in the listening condition was higher in the STG (mean = 0.56) than in the IFG (mean = 0.54) and SMC (mean = 0.54; $p < 0.05$; unpaired two-sample t-test; FDR correction). Finally, no significant differences across gyri were observed in the imagined speech condition ($p > 0.5$; unpaired two-sample t-test; Bonferroni correction).

3 Discussion

Our results provide the first demonstration of single-trial neural decoding of words during imagined speech production. We developed a new binary classification approach that accounted for temporal variations in the high frequency neural activity across speech utterances. We used support-vector machines to classify individual words in a word pair, and introduced a non-linear time alignment into the kernel to deal with internal speech production variability. At the group level, average classification accuracy across all pairs was significant in all three conditions. Two subjects that exhibited the lowest classification scores had right hemisphere coverage and were right handed, which is typically associated with left hemisphere language dominance [23]. This may have contributed to differences in accuracy across left and right hemisphere grid subjects. However, more data are required to delineate the effect of hemisphere coverage in the decoding process. The anatomic locations that led to the best word discrimination in the listening and overt speech conditions were consistent across subjects. All three anatomical regions (STG, IFG and SMC) provided information in the classification process. In the imagery condition, anatomical areas with the highest predictive power were more variable across subjects. The results revealed that the STG alone could drive auditory imagery decoding, but that other areas, such as the IFG and SMC also contribute.

An important component of the study is the application of dynamic time warping in the classification framework to account for speech production temporal irregularities. This technique maximizes alignment of the neural activity time courses without knowledge of the exact onset of the events. This approach proved useful for studying imagined speech where no behavior or stimuli are explicitly observed, and highlights the usefulness of a time alignment procedure such as using DTW for modeling the neural activity of unobserved behavioral events such as imagery. However, we also note the limitations of DTW in noisy environments, suggesting that imagery results may be improved by developing more robust realignment techniques. In addition, finding a behavioral or neural metric that allows marking more precisely the imagined speech onset and offset would reduce the temporal uncertainty window. This will be increasingly important when we move towards asynchronous protocols, i.e. when patients spontaneously produce imagined speech, as opposed to current protocols in which they are cued.

Despite intense investigation, it is still unclear how the content of imagined speech is processed in the human cortex. Different tasks—such as word repetition, letter or object naming, verb generation, reading, rhyme judgment, and counting—involve different speech production processes, ranging from lexical retrieval to phonological or even phonetic encoding [17]. In this study, we chose the set of auditory stimuli to maximize variability in several speech feature spaces (acoustic features, number of syllables, semantic categories), but to minimize word length variance. Our approach does not allow us to investigate which specific speech features provided information and allowed classification; i.e., if the discrimination was based on acoustic, phonetic, phonological, semantic or abstract features within speech perception, comprehension or production. Given that several brain areas were involved, it is likely that various features of speech were involved in the classification process.

Several additional limitations precluded high word prediction accuracy during imagined speech. First, we were limited by the electrode location and duration of implantation that was not designed for the experiments, but solely for clinical needs. Higher density grids placed at specific locations in the posterior superior temporal gyrus, frontal cortex and/or sensorimotor cortex that are active during imagined speech would provide higher spatial resolution and potentially enhanced discriminating signals. Further, subjects were not familiarized with the task beforehand (i.e. no training), and due to time constraints in the epilepsy-monitoring unit, we were unable to monitor subjects' performance or vividness during speech imagery. We also could not reject pronunciation and grammatical mistakes, as we did in the overt speech condition. We propose it would be beneficial to train subjects on speech imagery prior to surgery to enhance task performance. Finally, our current approach was limited in the set of choices available, and only tests binary classification between word pairs. In addition, the effect size is small, and likely not clinically significant for a communication interface. Classification of individual words among multiple other words or continuous speech decoding would be a more realistic clinical scenario. An alternative would be classifying phonemes, which forms the building blocks of speech instances. Decoding vowels and consonants in overt and imagined words using electrocorticographic signals in humans has shown promising results [3, 7], and would allow generating a larger lexicon from a fewer number of classes (60–80 phonemes in spoken English [25]).

Although the feasibility for closed-loop brain-computer interface (BCI) systems remains to be demonstrated, these data represent a proof of concept study for basic direct decoding of speech imagery, which opens the door to new communication interfaces that may allow for more natural speech-like communication in patients with severe communication deficits.

Acknowledgements This work was supported by the NIH (EB00856, EB006356 and EB018783), the US Army Research Office (W911NF-08-1-0216, W911NF-12-1-0109, W911NF-14-1-0440), Fondazione Neurone, Zeno-Karl Schindler Foundation, NINDS Grant R3721135 and the Nielsen Corporation, and is adapted with permissions from [12].

References

1. A. Aleman, The functional neuroanatomy of metrical stress evaluation of perceived and imagined spoken words. Cereb. Cortex **15**, 221–228 (2004). https://doi.org/10.1093/cercor/bhh124
2. L. Aziz-Zadeh, L. Cattaneo, M. Rochat, G. Rizzolatti, Covert speech arrest induced by rTMS over both motor and nonmotor left hemisphere frontal sites. J. Cogn. Neurosci. **17**, 928–938 (2005). https://doi.org/10.1162/0898929054021157
3. J.S. Brumberg, E.J. Wright, D.S. Andreasen et al., Classification of intended phoneme production from chronic intracortical microelectrode recordings in speech-motor cortex. Front. Neurosci. (2011). https://doi.org/10.3389/fnins.2011.00065
4. S. Geva, M. Correia, E.A. Warburton, Diffusion tensor imaging in the study of language and aphasia. Aphasiology **25**, 543–558. (2011). https://doi.org/10.1080/02687038.2010.534803
5. M. Gönen, A. Ethem, Multiple kernel learning algorithms. J. Mach. Learn. Res. 2211–2268 (2011)
6. T. Hastie, *The Elements of Statistical Learning: Data Mining, Inference, and Prediction*, 2nd edn. (Springer, New York, NY, 2009)
7. C. Herff, D. Heger, A. de Pesters et al., Brain-to-text: decoding spoken phrases from phone representations in the brain. Front Neurosci. (2015). https://doi.org/10.3389/fnins.2015.00217
8. R.M. Hinke, X. Hu, A.E. Stillman et al., Functional magnetic resonance imaging of Broca's area during internal speech. NeuroReport **4**, 675–678 (1993)
9. J. Huang, T.H. Carr, Y. Cao, Comparing cortical activations for silent and overt speech using event-related fMRI. Hum. Brain Mapp. **15**, 39–53 (2002)
10. J. Kubanek, P. Brunner, A. Gunduz et al., The tracking of speech envelope in the human cortex. PLoS ONE **8**, e53398 (2013). https://doi.org/10.1371/journal.pone.0053398
11. S. Martin, P. Brunner, C. Holdgraf et al., Decoding spectrotemporal features of overt and covert speech from the human cortex. Front. Neuroengineering **7**, 14 (2014). https://doi.org/10.3389/fneng.2014.00014
12. S. Martin, P. Brunner, I. Iturrate et al., Word pair classification during imagined speech using direct brain recordings. Sci. Rep. **6**, 25803 (2016). https://doi.org/10.1038/srep25803
13. P.K. McGuire, D.A. Silbersweig, R.M. Murray et al., Functional anatomy of inner speech and auditory verbal imagery. Psychol. Med. **26**, 29–38 (1996)
14. N. Mesgarani, E.F. Chang, Selective cortical representation of attended speaker in multi-talker speech perception. Nature **485**, 233–236 (2012). https://doi.org/10.1038/nature11020
15. E.D. Palmer, H.J. Rosen, J.G. Ojemann et al., An event-related fMRI study of overt and covert word stem completion. NeuroImage **14**, 182–193 (2001). https://doi.org/10.1006/nimg.2001.0779
16. B.N. Pasley, S.V. David, N. Mesgarani et al., Reconstructing speech from human auditory cortex. PLoS Biol. **10**, e1001251 (2012). https://doi.org/10.1371/journal.pbio.1001251
17. M. Perrone-Bertolotti, L. Rapin, J.-P. Lachaux et al., What is that little voice inside my head? Inner speech phenomenology, its role in cognitive performance, and its relation to self-monitoring. Behav. Brain Res. **261**, 220–239 (2014). https://doi.org/10.1016/j.bbr.2013.12.034
18. C.J. Price, A review and synthesis of the first 20 years of PET and fMRI studies of heard speech, spoken language and reading. NeuroImage **62**, 816–847 (2012). https://doi.org/10.1016/j.neuroimage.2012.04.062
19. L.R. Rabiner, *Fundamentals of Speech Recognition* (PTR Prentice Hall, Englewood Cliffs, N.J., 1993)

20. A. Ritaccio, P. Brunner, A. Gunduz et al., Proceedings of the fifth international workshop on advances in electrocorticography. Epilepsy Behav. **41**, 183–192 (2014). https://doi.org/10.1016/j.yebeh.2014.09.015

21. H. Sakoe, S. Chiba, Dynamic programming algorithm optimization for spoken word recognition. IEEE Trans. Acoust. Speech Signal Process. **26**, 43–49 (1978). https://doi.org/10.1109/TASSP.1978.1163055

22. S.S. Shergill, E.T. Bullmore, M.J. Brammer et al., A functional study of auditory verbal imagery. Psychol. Med. **31**, 241–253 (2001)

23. A.W. Toga, P.M. Thompson, Mapping brain asymmetry. Nat. Rev. Neurosci. **4**, 37–48 (2003). https://doi.org/10.1038/nrn1009

24. V.L. Towle, H.-A. Yoon, M. Castelle et al., ECoG gamma activity during a language task: differentiating expressive and receptive speech areas. Brain **131**, 2013–2027 (2008). https://doi.org/10.1093/brain/awn147

25. S.V. Vaseghi, *Multimedia Signal Processing: Theory and Applications in Speech, Music and Communications* (Wiley, Chichester, England ; Hoboken, NJ, 2007)

26. F.Z. Yetkin, T.A. Hammeke, S.J. Swanson et al., A comparison of functional MR activation patterns during silent and audible language tasks. AJNR Am. J. Neuroradiol. **16**, 1087–1092 (1995)

High Performance BCI in Controlling an Avatar Using the Missing Hand Representation in Long Term Amputees

Ori Cohen, Dana Doron, Moshe Koppel, Rafael Malach
and Doron Friedman

Abstract Brain-computer interfaces (BCIs) have been employed to provide different patient groups with communication and control that does not require the use of limbs that have been damaged. In this study, we explored BCI-based navigation in three long term amputees. Each participant attempted motor execution with the affected limb, and performed motor execution with the intact limb, while fMRI activity was recorded. Participants attempted, and executed, one of four tasks to direct the movement of an avatar on a monitor. Classification accuracy was very high across both cue-based and free-choice conditions. Results support the use of this fMRI BCI approach for virtual navigation, which could improve BCIs based on fMRI as well as other approaches such as EEG.

Keywords BCI · Long term amputees · Avatar · fMRI · SVM

O. Cohen · D. Friedman (✉)
Advanced Reality Lab, Interdisciplinary Center Herzliya (IDC H.), P.O. Box 167, Herzliya 46150, Israel
e-mail: doronf@idc.ac.il

O. Cohen
e-mail: orioric@gmail.com

O. Cohen · M. Koppel
Department of Computer Science, Bar-Ilan University, Ramat-Gan 52900, Israel

D. Doron
Department of Brain Injury Rehabilitation, Sheba Medical Center, Tel-Hashomer, Ramat-Gan, Israel

R. Malach
Department of Neurobiology, Weizmann Institute of Science, Rehovot 76100, Israel

© The Author(s), under exclusive licence to Springer Nature Switzerland AG 2019   93
C. Guger et al. (eds.), *Brain-Computer Interface Research*,
SpringerBriefs in Electrical and Computer Engineering,
https://doi.org/10.1007/978-3-030-05668-1_9

1   Introduction

Brain-computer interfaces (BCIs) are especially important for those otherwise inca-
pable of using their bodies to execute their intentions [1, 2]. For patient populations,
we need to address a major open question: can motor brain circuits that have been
deprived of input following trauma still be used for controlling a BCI? In cases of
long term amputation, both efferent and afferent functions are abolished and may
lead to deterioration of the relevant brain representations. Our study addresses this
question by allowing a group of long-term amputees to control a computer generated
avatar in real-time using their missing hand.

   We have previously demonstrated that functional magnetic resonance imaging
(fMRI), despite its dependence on sluggish hemodynamic signals, is capable of
delivering BCI control and allows subjects to perform complex navigation tasks [3]
or teleoperate a humanoid robot [4, 5]. FMRI offers advantages of being risk-free,
non-invasive, offering superior spatial resolution, and allowing us to tap the rich,
functionally complex cortical organization of the whole brain—properties that are
not matched by any other real time method currently available, including invasive
methods. Thus, we suggest that real time fMRI, even though it is not a target platform
for end user BCI by itself, can play a larger role in pushing the capabilities of novel
invasive and non-invasive BCIs.

   For patient populations, such as amputees, there is no need to use motor imagery.
Therefore, we asked our subjects to attempt motor execution using their missing limb,
compared with actual motor execution using the fingers of their intact hand and toes.
FMRI provides a highly detailed anatomical mapping of the brain activity, which
allows us a quantitative comparison of the brain areas responsible for healthy and
amputated limb control. The flexibility of our whole-brain machine learning-based
approach [6] enabled the participation of neuronal populations in missing-hand motor
actions, including voxels that are not typically associated with motor imagination or
motor attempt.

2   Materials

Imaging was performed on a 3T Trio Magnetom Siemens scanner as described in
[3], with a repetition time (TR) of 2000 ms. We used our system based on whole
brain machine learning [6] for training and real-time classification. Visual feedback
was provided by a mirror, placed 11 cm from the eyes of the subject and 97.5 cm
from a screen, which resulted in a total distance of 108.5 cm from the screen to the
eyes of the subject.

   In order to verify that the amputee subjects are not using their stumps, we connected electromyogram (EMG) electrodes to each subject's muscle area surrounding

the stump. Subjects were instructed to move the fingers in the amputated arm, and data (bandpass 1–5000 Hz; sampling rate 10,000 Hz) obtained from the shoulder area were collected to validate that no muscle activity was involved in motor movement of the amputated arm. A comb band stop filter was used with a fixed value of 16 Hz to remove repetitive noise that came from scanning 32 slices every 2000 ms.

3 Methods

Subjects: Seven subjects took part in the study: four control (2 male, mean age 28.5) and three amputees, all male (mean age 31.3), as follows: BZ is 40 years old, amputation above the elbow, 2 years after the accident. PW (Fig. 1) is 26 years old, amputation below the shoulder, 1.5 years after the accident. BH is 28 years old, amputation below the shoulder, 2 years after the accident. All subjects reported suffering from mild to high levels of phantom pain. Each subject performed multiple sessions over consecutive days.

Procedure: In the first part of this experiment, the subject sees an avatar standing in the center of a room (Fig. 2a). In each trial, the subject is given 40 pseudo-random auditory instructions ("left", "right", "forward", and "rest"), 10 from each class,

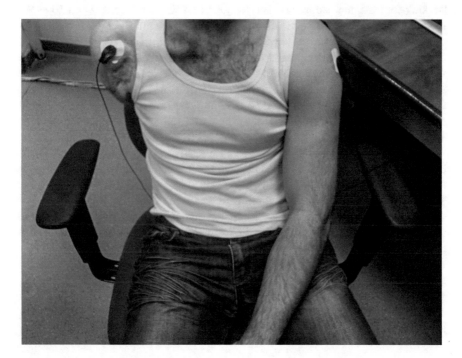

Fig. 1 Amputee subject PW with the EMG electrodes connected to the shoulder

(a)   **(b)**

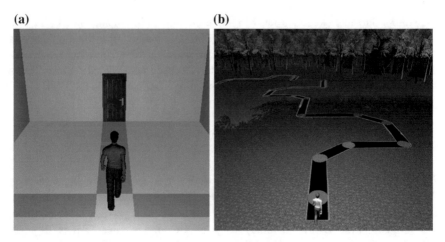

Fig. 2 Snapshots of the virtual environment displayed on the monitor during the study. **a** An avatar standing in the center of a room. **b** The 3D virtual path scenario. The subject's avatar is seen standing at the beginning of the path

based on a motor execution (or motor attempt) experimental protocol. Six seconds after each action, the subject is instructed to rest. During that time the avatar executes the pre-determined command that corresponds to the instruction (turning left or right, walking forward, and stopping). The rest duration varies between 8 and 10 s. We record between 3 and 4 sessions as input to a machine-learning tool [6]. For purposes of learning, we select only those voxels with highest information gain (IG). Labeled training examples are then passed on to our learning algorithm. The result of the training phase is a support vector machine (SVM) model that can classify previously unseen vectors. In the second part, the subjects perform a free-choice navigation task. Each subject was instructed to guide an avatar toward the end of a path by picking up as many discs as possible (Fig. 2b). The avatar must "touch" a disc in order to successfully collect it, and then the disc changes to green. Our system classifies the subjects' intentions every TR (in our case 2 s) in real time. Selecting the same voxels based on the IG filtering performed at model training, the data are passed into the trained SVM model, and the classification result is transmitted to the avatar (we use the Unity 3D engine for virtual environment feedback). Each trial lasted 696 s.

Results: Classification accuracy in the cue-based BCI task (Fig. 3) and in the free-choice task (Fig. 4) is promising, indicating that the degree of BCI control and performance with a missing limb is very high and comparable to the performance with the intact hand, and to the performance of control participants. In the free-choice task, the amputees showed similar trajectory patterns to the control group (Fig. 5), and their performance was only slightly lower than controls (Fig. 4).

A mixed effects for repeated measures statistical analysis taking into account subject, condition and accuracy indicated no significant difference between the groups in

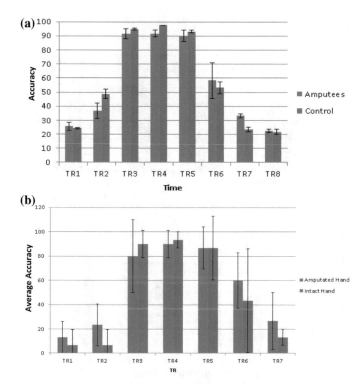

Fig. 3 **a** Average classification accuracy in each TR for both groups. Accuracy was calculated from four available classes, with a chance level of 25%. **b** Average classification accuracy in each TR for amputated and intact hands. Error bars indicate the 95% confidence interval

TR3 (amputees $= 91.6\%$, control $= 95\%$, $p = 0.45$) and higher performance (nearly significant) in controls at TR4 (amputees $= 91.6\%$, control $= 97.5\%$, $p = 0.068$) (Fig. 3a). Figure 3b shows similar average classification accuracy in the amputee group for the intact- and missing-hand (3 subjects), taking into account the TR with maximum classification. Figure 4a shows average command usage in the free-choice trials, indicating similar usage patterns; i.e., the amputees had no bias toward the intact hand. A mixed effects for repeated measures statistical analysis taking into account subject, condition and performance, indicated that the control group performed slightly better in the free choice task, but the difference was not significant ($p = 0.158$) (Fig. 4b). A significant difference was found among the subjects ($p = 0.024$).

Both subject populations were able to adopt a strategy that evoked motor-related brain regions. Nevertheless, an analysis of the voxels selected by the machine-learning platform for classification indicates larger variability in the amputee population and in some cases, there is evidence that the motor activation extended beyond the motor regions, as seen in Fig. 6.

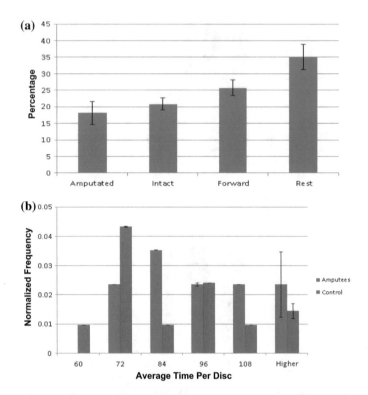

Fig. 4 **a** Class usage percentage for the amputated group for each command out of 100%. **b** The distributions of the number of discs collected per group, based on the average time to collect a single disc. The Y-axis represents the normalized frequency and the X-axis represents each increment. Error bars indicate the 95% confidence interval

Conclusion: In our study [7, 8], we used real-time fMRI to demonstrate remarkably high performance of long term amputees in controlling a virtual avatar using their missing hand. Our whole-brain machine learning system was able to select the most relevant patterns for the task, without prior assumptions or information about those brain regions, converging on motor areas, for all subjects. We also demonstrate the utility of real-time fMRI for BCI: fMRI offers advantages of anatomical detail and brain coverage that are not matched by any other real time method currently available, including invasive methods. Thus, we suggest the fMRI-based BCI can make a great contribution to BCI development, as well as to individual training and adaptation. fMRI-based BCI can be used to develop algorithms tailored to individuals following brain reorganization, and the algorithms can adapt to further neural changes following BCI training. These issues are crucial for clinical populations.

Another issue of interest for future research is the ideal training paradigm. Details such as the frequency and duration of BCI usage sessions, type of avatar and how it interacts with the user, and the best instructions to the end-user all merit further

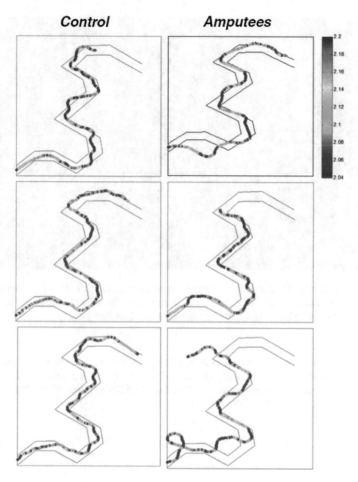

Fig. 5 A visualization of the paths of the best performance navigation trial of six subjects. Left column: controls, right column: amputees. Warmer colors reflect higher speed

study. For example, we asked subjects to attempt movement rather than imagine movement. Many BCIs that utilize movement activity instruct subjects to imagine movement [1, 2, 3, 9, 10]. The instructions to the end-user can influence motor activity and BCI classification [11], but this issue has not been well explored with patient populations. Studies should explore different instructions such as attempting versus imagining movement, with consideration of not only which mechanism yields activity that is easier to detect, but also which approach that patients prefer.

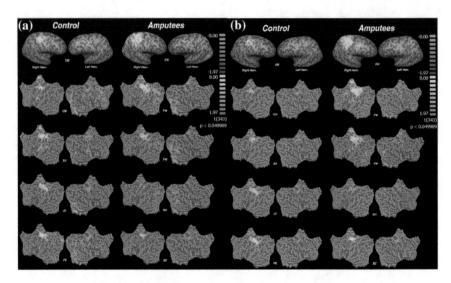

Fig. 6 A gallery visualization of the left versus right contrast using a $p < 0.05$ (uncorrected) for the **a** cue-based and **b** free-choice tasks, for all subjects from the amputees and the control group. The red arrows represent the dominant hand and intact hand for each patient

Acknowledgements This research was supported by the European Union FP7 Integrated Project VERE (No. 657295), http://www.vereproject.eu. We would like to thank the subjects for their participation, and the Weizmann Institute fMRI scanner staff Edna Furman-Haran, Nachum Stern and Fanny Attar for their help in this experiment.

References

1. J. Wolpaw, E.W. Wolpaw (eds.), *Brain-Computer Interfaces: Principles and Practice* (OUP, USA, 2012)
2. C.S. Nam, A. Nijholt, F. Lotte (eds.), *Brain–Computer Interfaces Handbook: Technological and Theoretical Advances* (CRC Press, 2018)
3. O. Cohen, M. Koppel, R. Malach, D. Friedman, Controlling an avatar by thought using real-time fMRI. J. Neural Eng. **11**(3), 035006 (2014)
4. P. Andersson, J.P. Pluim, M.A. Viergever, N.F. Ramsey, Navigation of a telepresence robot via covert visuospatial attention and real-time fMRI. Brain Topogr. **26**(1), 177–85 (2013). https://doi.org/10.1007/s10548-012-0252-z
5. O. Cohen, S. Druon, S. Lengagne, A. Mendelsohn, R. Malach, A. Kheddar, D. Friedman, fMRI-based robotic embodiment: controlling a humanoid robot by thought using real-time fMRI. Presence: Teleoperators and Virtual Environ. **23** (3), 229–241 (2014)
6. O. Cohen, M. Ramot, R. Malach, M. Koppel, D. Friedman, A generic machine-learning tool for online whole brain classification from fMRI, in *The 6th International Brain-Computer Interface Conference*, Graz, Austria (2014)

7. O. Cohen, R. Malach, M. Koppel, D. Friedman, Can amputees control a brain-computer interface with their missing hand? in *Proceedings of the Sixth International Brain-Computer Interface Meeting: BCI Past, Present, and Future* (Asilomar, California, 2016), p. 107

8. O. Cohen, D. Doron, M. Koppel, R. Malach, D. Friedman, High performance in brain-computer interface control of an avatar using the missing hand representation in long term amputees, in *The 8th International IEEE EMBS Conference On Neural Engineering (NER17)*, Shanghai, China (2017)

9. D.C. Irimia, W. Cho, R. Ortner, B.Z. Allison, B.E. Ignat, G. Edlinger, C. Guger, Brain-computer interfaces with multi-sensory feedback for stroke rehabilitation: a case study. Artif. Organs **41**(11), E178–E184 (2017). https://doi.org/10.1111/aor.13054

10. L. Acqualagna, L. Botrel, C. Vidaurre, A. Kübler, B. Blankertz, Large-scale assessment of a fully automatic co-adaptive motor imagery-based brain computer interface. PLoS ONE **11**(2), e0148886 (2016). https://doi.org/10.1371/journal.pone.0148886

11. C. Neuper, R. Scherer, M. Reiner, G. Pfurtscheller, Imagery of motor actions: differential effects of kinesthetic and visual-motor mode of imagery in single-trial EEG. Brain Res. Cogn. Brain Res. **25**(3), 668–77 (2005)

Can BCI Paradigms Induce Feelings of Agency and Responsibility Over Movements?

Birgit Nierula and Maria V. Sanchez-Vives

Abstract The sense of agency is the attribution of an action to ourselves, which allows us to distinguish our own actions from those of other people and gives us a feeling of control and responsibility for their outcomes. Under physiological conditions, the sense of agency typically accompanies all our actions. Further, it can even be experienced over an illusory owned body—that is, a surrogate body perceived as if it were our own. However, the extent to which actions controlled through a brain–computer interface (BCI) also induce feelings of agency and responsibility is not well known. In the following chapter, we will review the relevant literature on body ownership and agency in virtual reality (VR) embodiment and outline an experiment in which participants controlled a virtual body through different BCI protocols based either on sensorimotor activity or on visually evoked potentials. Our findings show that BCI protocols can induce feelings of agency and that those BCI protocols based on sensorimotor activity have an advantage over those based on activity in visual areas. We further show that BCI protocols based on sensorimotor activity can even induce feelings of responsibility over the outcomes of that action, a finding that raises important ethical implications. We give particular focus to subjective reports from the debriefing after the experiment about the experience of BCI-induced agency over the action of a virtual body.

Keywords BCI · Agency · Immersion · EEG · Avatar · Virtual reality · Virtual embodiment

B. Nierula · M. V. Sanchez-Vives (✉)
Institut d'Investigacions Biomèdiques August Pi i Sunyer (IDIBAPS), Barcelona, Spain
e-mail: sanchez.vives@gmail.com

Event-Lab, Department of Clinical Psychology and Psychobiology, Universitat de Barcelona, Barcelona, Spain

M. V. Sanchez-Vives
Institució Catalana Recerca i Estudis Avançats (ICREA), Barcelona, Spain

Departamento de Psicología Básica, Universitat de Barcelona, Barcelona, Spain

B. Nierula
Max Planck Institute for Cognitive and Brain Sciences, Leipzig, Germany

1 How to Move an Embodied Virtual Body

Body ownership illusions allow subjects to *embody* a surrogate body such that they perceive this body to be their own. These illusions can be induced not only over physical bodies like mannequins [25] but also over virtual ones [32]. Body ownership illusions persist even if the embodied bodies have an appearance that is different from the real one—such as that of a child [1], or of a person with a different race [23], or even that of a robot [14]. Body ownership can be induced through synchronous multisensory correlations, like visuotactile [3, 25, 31] or visuomotor correlations [17, 30]. To induce the latter in a virtual environment, the surrogate body is moved synchronously with the person's real movements—this is achieved by capturing the person's movements with a tracking system and by mapping them onto the virtual body. Another possible way to move a virtual body is through brain–computer interfaces (BCIs). BCIs allow direct communication from a participant's brain to a computer, thus circumventing the physiological pathway, which evolves from an intention through a body action to an outcome, by connecting intention and outcome through a feedback loop. BCIs have already been used to control events in virtual environments [6, 8, 24, 27]. One of their advantages is that, using a BCI, one can generate goal-directed actions "by thinking", in what could be considered a more direct (if not natural) way to control an action than, for example, using a joystick to generate the same action [5, 11, 37].

2 The Sense of Agency of Our Own Movements

Our sense of agency accompanies actions that we attribute to ourselves and thereby gives us the subjective sensation that we are in control of our own actions (for a review see [4, 9]); it further creates the basis for our feeling of responsibility over the outcome of our actions [10]. It is however not clear the extent to which BCI-controlled actions can also induce agency or feelings of responsibility over surrogate, either virtual or robotic, bodies. This is a highly relevant issue (1) for the participant's user-experience, and (2) for the important ethical implications it entails—for example, does one feel responsible when controlling a remotely located robot through BCI? In this chapter, we briefly review relevant literature on agency and we highlight the importance of body ownership for experiencing illusions of agency. We finish the chapter with a description of an experiment in which participants controlled the movement of an embodied virtual body through two different BCI paradigms: sensorimotor rhythm-based motor imagery and steady-state visual-evoked potential (SSVEP).

Agency contributes to the distinction between self- and externally generated actions and their effects. Voluntary actions always start with an intention, after which there is an action selection process and a motor plan, which then transform the intention into an action. An internal comparator process then links the action's outcome to its motor plan and its intention by comparing the actual with the predicted sensory feedback (based on the motor plan) and the intention [41]. Such internal comparisons are believed to be the basis of motor learning because they allow the adjustment and improvement of the motor plan and its feedback predictions, but they also allow the identification of errors and the distinguishing of external from internal actions. In fact, the comparator mechanism between expected and actual sensory feedback has also been proposed to result in the sense of agency [7].

3 Brain Signatures of Movement Errors

In a recent electroencephalography (EEG) study using virtual reality, Padrao et al. [22] identified two error-monitoring loops involved in the generation of the sense of agency on a neurophysiological level. In this study, participants were embodied in a virtual body and performed a standard fast reaction time task, in which in certain trials the avatar movement did not match the participant's movement. Therefore, the task produced self-generated real errors and avatar-generated false errors. While real errors showed an expected error-related negativity (Ne/ERN) around 100 ms after the error over fronto-central electrodes, avatar-generated false errors produced a delayed negative response around 300–400 ms after the error over parietal regions (Fig. 1), similar to the N400, which is an event-related potential (ERP) component related to semantic or conceptual violations. This false error-related component was not elicited when participants saw another avatar performing the same task from a third-person perspective.

While the comparator mechanism is widely accepted as the correct explanation of the sense of agency, several studies have shown that the sense of agency is also driven by cognitive processes [33]. For example, the level of cognitive preparation [12] and the process of selecting between alternative actions [13, 40] contribute to the sense of agency. In addition, Synofzik et al. [34] proposed a two-step account of agency, which assumes the production of a basic, perceptual representation of agency, which is further processed on a cognitive level.

4 Agency and Body Ownership

Agency is also closely related to the sense of body ownership, which is the feeling that a body is one's own body. In healthy humans, body ownership and agency contribute to a coherent perception of body awareness. In some psychiatric conditions, however, the sense of agency can be separated from the sense of body ownership; for example, patients with schizophrenia who experience delusions of control can believe that another authority is controlling the actions of their own body. Illusions of body ownership over virtual body parts and full virtual bodies are typically induced through simultaneous visuotactile [3, 25, 31] or visuomotor [30] stimulation. A study by Perez-Marcos et al. [24] showed that it is even possible to induce body ownership

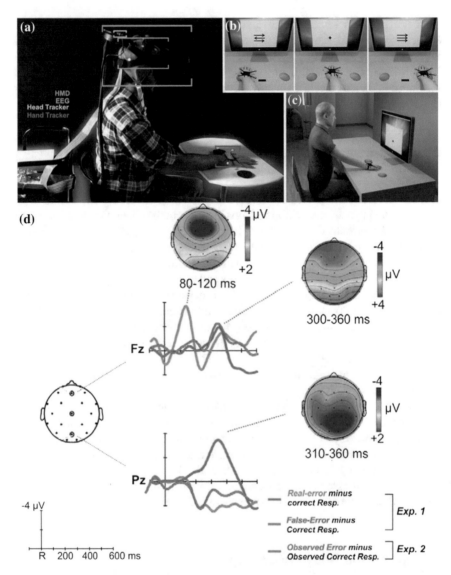

Fig. 1 **a** Participant wearing EEG and a head-mounted display through which they saw the virtual environment. **b** Participant's view of the virtual scenario from a first-person perspective and **c** from a third-person perspective. **d** Event-related responses to observed errors over which participants either felt agency (real errors) or not (false errors and observed errors). Participants made real errors (blue, performed by the participant) and false errors (green, performed by the avatar) while they were seeing the actions from a first-person perspective. Instead, observed errors (red, performed by an avatar) were seen from a third-person perspective by observing an avatar performing the task. All graphs show the difference in waveforms and related scalp distribution maps to the correct response. Real errors showed the frontocentral error-related negativity (ERN) peaking at about 100 ms, while false errors yielded a slower negative parietal component (at around 300–400 ms). Figure reproduced with permission from Padrao et al. [22]

Fig. 2 Controlling the movements of a virtual arm through a motor imagery-based BCI can induce intermediate levels of body ownership over a virtual body. Figure reproduced with permission from Perez-Marcos et al. [24]

through a BCI. In this study, participants controlled by thought the opening and closing of a virtual hand placed at the end of an arm coming out of the shoulder of the participant. Through a motor imagery task, the participants controlled the hand movements, which along with the feasible position and first-person point of view resulted in an illusion of ownership of the virtual arm (Fig. 2).

Agency has been argued to be only experienced over voluntary movements, while body ownership is always present at rest, and during passive and voluntary movements [16, 36]. A recent experiment performed in virtual reality, however, showed that when people experience high levels of body ownership over a virtual body, they can perceive an action such as the speaking of the virtual avatar as if it was their own [2]. In this experiment, participants saw a virtual body from a first-person perspective that was co-located with their real body. They performed a set of exercises in which the virtual body was either moving synchronously or not with their real body, which led to either high or low levels of body ownership, respectively (Fig. 3). Then, participants were asked to focus on their virtual body and face seen in a mirror in the virtual environment while occasionally performing some exercises. During this time, the virtual avatar was speaking with a fundamental frequency higher than their own voice. Participants who experienced high levels of body ownership in the synchronous movement condition attributed the speaking of the avatar to themselves and changed the fundamental frequency of their voice towards the fundamental frequency of the avatar. This was not the case for people with low body ownership. Half of the participants additionally felt vibrations on their larynx when the avatar was speaking. This further attributed to the effect in the way that those participants with additional vibrations tended to higher feelings of body ownership and agency; however, it could not explain the self-attribution of the avatar's speaking by itself.

Further evidence in the same direction was provided by Kokkinara et al. [15], who showed that when participants experience body ownership over a virtual body, they attribute the walking of that virtual body to themselves, although they were sitting on a chair. These studies show that strong body ownership over a surrogate body plays

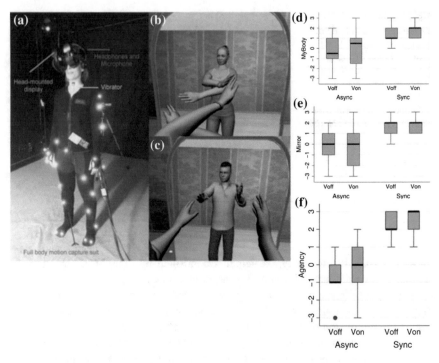

Fig. 3 a Participant wearing a head-mounted display and a tracking suit. **b** Virtual scenario seen from a first-person perspective of the female **c** and the male avatars. Participants could see the virtual body directly when they looked down or moved their arms in front of their body, or when they looked in the mirror. **d, e** Feeling of body ownership over the virtual body when participants looked down on themselves (MyBody) and for the body they saw when they looked in the virtual mirror (Mirror) were higher during synchronous (Sync) than during asynchronous visuomotor stimulation (Async). **f** Agency ratings were higher during Sync than during Async. (Voff = without vibrations at the thyroid; Von = with vibrations at the thyroid). Figure reproduced with permission from Banakou and Slater [2]

a crucial role when attributing its actions to oneself. Therefore, motor preparation is not always necessary for a sense of agency.

5   Is There Agency in BCI-Controlled Movements?

BCIs are another example in which body movements are not directly controlled by brain activity but indirectly through a computer. By these means, brain activity is used to control actions, often through surrogates of the actual body: prostheses, robotic arms or virtual bodies. Different BCI protocols use different neural activity patterns; for example, in sensorimotor areas (sensorimotor rhythm-based BCIs), or sensory-evoked (e.g., steady-state visual-evoked potential (SSVEP) based BCIs), amongst others. Sensorimotor rhythms (SMR) based BCI protocols use motor imagery to produce changes in alpha- and/or beta-oscillations over somatosensory areas. These changes can be seen as a decrease in band power relative to a baseline, known as event-related desynchronization (ERD), or as an increase, known as event-related synchronization (ERS) [28]. Due to the somatotopic layout of the sensorimotor cortex, changes in SMR occur in different locations depending on the body part that is imagined moving and can therefore be extracted by a BCI algorithm. Alpha and/or beta-ERD can be observed during planning, imagination, observation, and execution of a movement [18, 19, 21, 26, 29]. SSVEP-based BCI protocols exploit visual stimuli that flicker at a specific frequency. When centered in the fovea, such stimuli produce oscillations at the simulation frequency over the visual cortex, which the BCI algorithm can identify with relatively high accuracy.

BCIs have been successfully used in combination with VR for over a decade [6, 8, 24, 27]. Because they couple an intention with its outcome, they should have the ability to induce a sense of agency over BCI-induced actions [9, 38]. This claim is supported by a recent pilot study, which aimed at replicating a study by Wegner et al. [39], in which participants felt agency over another person's hand movements by means of BCI. However, no conclusions could be drawn from the data due to the small sample size [35].

In a recent study by Nierula et al. [20], we addressed the following two questions: (1) is there a BCI paradigm that has a greater likelihood to induce agency and responsibility over BCI-controlled movements? And (2) is it necessary to activate motor areas in order to induce agency and responsibility over a BCI-controlled movement? By answering these questions, we aimed at (1) obtaining information about the brain basis of agency and (2) exploiting a practical application of this research: if inducing agency and responsibility over movements is relevant to a particular application of BCI-controlled movement, which BCI paradigm is better to use?

6   Which BCI Paradigm Is Better to Induce Agency and Responsibility Over Movements?

In a study by Nierula et al. [20], an illusion of ownership of a virtual body was induced. Next, participants had to carry out a task with their arm. In order to control the movement of their "own" virtual arm (Fig. 4a, b) they used a BCI that exploited activity in (1) sensorimotor areas (motor imagery, MI) or (2) visual areas (SSVEP) or (3) a control condition, in which participants passively observed (OBS) the movements of the virtual body. All subjects performed all three conditions, which were presented in a random order. The conditions were: (1) controlling a virtual arm by imagining the arm movement (MI); (2) controlling the virtual arm by looking at a blinking button (SSVEP); and (3) observing the virtual arm moving without controlling it (OBS).

Fig. 4 a Position of participant during the experiment wearing a 59-channel EEG cap, a head-mounted display, and headphones. **b** The virtual environment from a first-person perspective as seen by the participant

The experience of control is part of the sense of agency. Participants gave highest ratings of control in the MI condition, while SSVEP came second. The lowest ratings were detected in OBS conditions [20]. Our data show that BCI protocols can induce higher levels of control than passively observing movement, which is probably due to their ability to connect intentions with the action. MI, however, induced the highest levels of control, highlighting the contributions of motor areas in the experience of control. Interestingly, when asked how responsible participants felt for an action that was accidentally performed by the virtual arm, they reported only high feelings of responsibility in the MI condition, while SSVEP and OBS conditions were similarly low [20]. We were further interested in participants' subjective experience of controlling a virtual body through a BCI. In the following, we give a short overview over the individual responses given by participants after the experiment by Nierula et al. [20] in an open response format. For some participants, the experience of illusory agency over the virtual body felt strange at first, but as the experimental condition proceeded, they increasingly perceived the virtual arm to be part of their body. For other participants, the visual attention directed to the blinking button during the SSVEP condition seemed to play an important role. Those participants reported that they felt stronger body ownership over the virtual body during the OBS condition compared with the SSVEP condition, because their focus on the button made them less aware of the virtual body. Other participants reported they felt less responsible for the actions of the virtual body when the BCI could not classify their brain activity in the previous trial and they therefore were not able to control the virtual arm. Two out of 29 participants reported the strongest feeling of agency during the OBS condition. Interestingly, these participants also reported that they felt more relaxed during that condition and therefore could allow the feeling of body ownership and agency to arise; instead, during the BCI conditions, they felt very tense and therefore felt less agency over the virtual body. Three participants felt the strongest feeling of agency during the SSVEP condition, and some of those reported that the imagination of the movement during the MI condition did not always coincide with the virtual movement. This feeling was particularly pronounced when the virtual arm started moving towards the button while they imagined the arm to be moving back to the resting position. We also asked participants how they imagined the movement. Some reported that they only imagined the movement towards the button, whilst others also imagined the return movement. Others reported that they felt less connected to the virtual body when imagining the movement more than once. Some participants reported that they were sure they imagined the movement in the right way, but the arm simply did not move as expected and this reduced their feeling of agency over the virtual movement. Such subjective reports should be taken into consideration when designing BCI applications that aim to induce high levels of agency over the movements of a surrogate body.

7 Conclusions

In fully immersive virtual environments in which we have a co-located body that is seen from a first-person perspective, it is possible not only to feel ownership over the virtual body but also to feel agency and control over the movements of this avatar body. When the movements of this avatar are controlled using a BCI, we found that BCI paradigms exploiting the activation of sensorimotor areas result in a higher agency over the controlled movements as well as responsibility over the actions. This should be taken into account by BCI creators that want to use current (or generate novel) paradigms of BCI for controlling movements locally or even remotely. There are further ethical concerns regarding how responsible subjects feel over mentally controlled actions. It is therefore beneficial to know that the sense of responsibility over actions is stronger when using motor imagery. We further provide new insights into participants' experiences. In order to feel agency over a virtual movement, it seems to be important that participants stay relaxed while using the BCI. The extent to which prolonged use of a different paradigm (e.g. SSVEP) might eventually induce agency remains to be explored.

References

1. D. Banakou, R. Groten, M. Slater, Illusory ownership of a virtual child body causes overestimation of object sizes and implicit attitude changes. Proc. Natl. Acad. Sci. U.S.A. **110**(31), 12846–12851 (2013)
2. D. Banakou, M. Slater, Body ownership causes illusory self-attribution of speaking and influences subsequent real speaking. Proc. Natl. Acad. Sci. U.S.A. **111**(49), 17678–17683 (2014). https://doi.org/10.1073/pnas.1414936111

3. M. Botvinick, J. Cohen, Rubber hands "feel" touch that eyes see. Nature **391**(6669), 756 (1998). https://doi.org/10.1038/35784
4. V. Chambon, N. Sidarus, P. Haggard, From action intentions to action effects: how does the sense of agency come about? Front. Hum. Neurosci. **8**, 320 (2014)
5. J.L. Collinger, B. Wodlinger, J.E. Downey, W. Wang, E.C. Tyler-Kabara, D.J. Weber, A.B. Schwartz, High-performance neuroprosthetic control by an individual with tetraplegia. The Lancet **381**(9866), 557–564 (2013). https://doi.org/10.1016/S01406736(12)61816-9
6. D. Friedman, R. Leeb, C. Guger, A. Steed, G. Pfurtscheller, M. Slater, Navigating virtual reality by thought: what is it like? Presence: Teleoperators and Virtual Environ. **16**(1), 100–110 (2007)
7. C.D. Frith, S.J. Blakemore, D.M. Wolpert. Abnormalities in the awareness and control of action. Philoso. Trans. Royal Soci. London. Series B: Biol. Sci. **355**(1404), 1771–1788 (2000). https://doi.org/10.1098/rstb.2000.0734
8. C. Guger, C. Groenegress, C. Holzner, G. Edlinger, M. Slater, Brain-computer interface for virtual reality control. Cyberpsychology & Behav. **12**(1), 84 (2009)
9. P. Haggard, Sense of agency in the human brain. Nat. Rev. Neurosci. **18**(4), 196–207 (2017). Retrieved from http://dx.doi.org/10.1038/nrn.2017.14
10. P. Haggard, M. Tsakiris, The experience of agency: feelings, judgments, and responsibility. Curr. Dir. Psychol. Sci. **18**(4), 242–246 (2009). https://doi.org/10.1111/j.1467-8721.2009.01644.x
11. L.R. Hochberg, D. Bacher, B. Jarosiewicz, N.Y. Masse, J.D. Simeral, J. Vogel, J.P. Donoghue, Reach and grasp by people with tetraplegia using a neurally controlled robotic arm. Nature **485**(7398), 372–375 (2012). https://doi.org/10.1038/nature11076
12. H.-G. Jo, M. Wittmann, T. Hinterberger, S. Schmidt, The readiness potential reflects intentional binding. Front. Hum. Neurosci. **8**, 421 (2014). https://doi.org/10.3389/fnhum.2014.00421
13. N. Khalighinejad, S. Di Costa, P. Haggard, Endogenous action selection processes in dorsolateral prefrontal cortex contribute to sense of agency: a meta-analysis of tDCS studies of "intentional binding". Brain Stimulation: Basic, Transl. Clin. Res. Neuromodulation **9**(3), 372–379 (2018). https://doi.org/10.1016/j.brs.2016.01.005
14. S. Kishore, M. Gonzalez-Franco, C. Hintermuller, C. Kapeller, C. Guger, M. Slater, K.J. Blom. Comparison of SSVEP BCI and eye tracking for controlling a humanoid robot in a social environment. Presence: Teleoperators and Virtual Environ. **23**(3), 242–252 (2014). https://doi.org/10.1162/PRES_a_00192
15. E. Kokkinara, K. Kilteni, K.J. Blom, M. Slater, First person perspective of seated participants over a walking virtual body leads to illusory agency over the walking. Sci. Rep. **6**, 28879 (2016). https://doi.org/10.1038/srep28879
16. M.R. Longo, P. Haggard, Sense of agency primes manual motor responses. Perception **38**(1), 69–78 (2009). https://doi.org/10.1068/p6045
17. A. Maselli, M. Slater, The building blocks of the full body ownership illusion. Front. Hum. Neurosci. **7**, 83 (2013). https://doi.org/10.3389/fnhum.2013.00083
18. C. Neuper, M. Wörtz, G. Pfurtscheller, ERD/ERS patterns reflecting sensorimotor activation and deactivation, in *Progress in Brain Research: Event-Related Dynamics of Brain Oscillations*, vol. 159, ed. by C. Neuper, W. Klimesch (Elsevier, Amsterdam, 2006), pp. 211–222. https://doi.org/10.1016/S0079-6123(06)59014-4
19. B. Nierula, F.U. Hohlefeld, G. Curio, V.V. Nikulin, No somatotopy of sensorimotor alpha-oscillation responses to differential finger stimulation. NeuroImage **76**, 294–303 (2013). https://doi.org/10.1016/j.neuroimage.2013.03.025
20. B. Nierula, B. Spanlang, M. Martini, M. Borrell V.V. Nikulin, M.V. Sanchez-Vives, Can we induce agency and responsibility over movements of a virtual body controlled through a brain computer interface? (2018) BiorXiv (in press)
21. V.V. Nikulin, F.U. Hohlefeld, A.M. Jacobs, G. Curio, Quasi-movements: a novel motor—cognitive phenomenon. Neuropsychologia **46**(2), 727–742 (2008). https://doi.org/10.1016/j.neuropsychologia.2007.10.008

22. G. Padrao, M. Gonzalez-Franco, M.V. Sanchez-Vives, M. Slater, A. Rodriguez-Fornells, Violating body movement semantics: neural signatures of self-generated and externalgenerated errors. NeuroImage **124**(Part A), 147–156 (2016). https://doi.org/10.1016/j.neuroimage.2015.08.022

23. T. Peck, S. Seinfeld, M. Aglioti, M. Slater, Putting yourself in the skin of a black avatar reduces implicit racial bias. Conscious. Cogn. **22**(3), 779–787 (2013). https://doi.org/10.1016/j.concog.2013.04.016

24. D. Perez-Marcos, M. Slater, M.V. Sanchez-Vives, Inducing a virtual hand ownership illusion through a brain-computer interface. NeuroReport **20**(6), 589–594 (2009). https://doi.org/10.1097/WNR.0b013e32832a0a2a

25. V.I. Petkova, H.H. Ehrsson, If I were you: perceptual illusion of body swapping. PLoS ONE **3**(12), e3832 (2008). https://doi.org/10.1371/journal.pone.0003832

26. G. Pfurtscheller, A. Aranibar, Evaluation of event-related desynchronization (ERD) preceding and following voluntary self-paced movement. Electroencephalogr. Clin. Neurophysiol. **46**(2), 138–146 (1979)

27. G. Pfurtscheller, R. Leeb, C. Keinrath, D. Friedman, C. Neuper, C. Guger, M. Slater, Walking from thought. Brain Res. **1071**(1), 145–152 (2006). https://doi.org/10.1016/j.brainres.2005.11.083

28. G. Pfurtscheller, F.H. Lopes da Silva, Event-related EEG/MEG synchronization and desynchronization: basic principles. Clin. Neurophysiol.: Official J. Int. Fed. Clin. Neurophysiol. **110**(11), 1842–1857 (1999)

29. G. Pfurtscheller, C. Neuper, Motor imagery activates primary sensorimotor area in humans. Neurosci. Lett. **239**(2–3), 65–68 (1997)

30. M.V. Sanchez-Vives, B. Spanlang, A. Frisoli, M. Bergamasco, M. Slater, Virtual hand illusion induced by visuomotor correlations. PLoS ONE **5**(4), e10381 (2010). https://doi.org/10.1371/journal.pone.0010381

31. M. Slater, D. Perez-Marcos, H.H. Ehrsson, M.V. Sanchez-Vives, Towards a digital body: the virtual arm illusion. Front. Hum. Neurosci. **2**, 6 (2008). https://doi.org/10.3389/neuro.09.006.2008

32. M. Slater, B. Spanlang, M.V. Sanchez-Vives, O. Blanke, First person experience of body transfer in virtual reality. PLoS ONE **5**(5), e10564 (2010). https://doi.org/10.1371/journal.pone.0010564

33. M. Synofzik, G. Vosgerau, A. Newen, Beyond the comparator model: a multifactorial two-step account of agency. Conscious. Cogn. **17**(1), 219–239 (2008). https://doi.org/10.1016/j.concog.2007.03.010

34. M. Synofzik, G. Vosgerau, A. Newen, I move, therefore I am: a new theoretical framework to investigate agency and ownership. Soc. Cognit. Emot. Self Conscious. **17**(2), 411–424 (2008). https://doi.org/10.1016/j.concog.2008.03.008

35. J.-P. van Acken, *Tracking the sense of agency in BCI applications* (Radboud University Nijmegen, Nijmegen, 2012)

36. E. van den Bos, M. Jeannerod, Sense of body and sense of action both contribute to selfrecognition. Cognition **85**(2), 177–187 (2002). https://doi.org/10.1016/S0010-0277(02)00100-2

37. M. Velliste, S. Perel, M.C. Spalding, A.S. Whitford, A.B. Schwartz, Cortical control of a prosthetic arm for self-feeding. Nature **453**, 1098–1101 (2008). https://doi.org/10.1038/nature06996

38. R. Vlek, J.-P. van Acken, E. Beursken, L. Roijendijk, P. Haselager, BCI and a user's judgment of agency, in *Brain-Computer-Interfaces in their Ethical, Social and Cultural Contexts*, vol. 12, ed. by G. Grübler, E. Hildt (Springer Netherlands, 2014), pp. 193–202. https://doi.org/10.1007/978-94-017-8996-7

39. D.M. Wegner, B. Sparrow, L. Winerman, Vicarious agency: experiencing control over the movements of others. J. Pers. Soc. Psychol. **86**(6), 838–848 (2004). https://doi.org/10.1037/0022-3514.86.6.838

40. D. Wenke, S.M. Fleming, P. Haggard, Subliminal priming of actions influences sense of control over effects of action. Cognition **115**(1), 26–38 (2010). https://doi.org/10.1016/j.cognition.2009.10.016
41. D.M. Wolpert, Z. Ghahramani, M.I. Jordan, An internal model for sensorimotor integration. Science **269**(5232), 1880 (1995). https://doi.org/10.1126/science.7569931

Recent Advances in Brain-Computer Interface Research—A Summary of the 2017 BCI Award and BCI Research Trends

Christoph Guger, Brendan Z. Allison and Natalie Mrachacz-Kersting

Abstract This book reviews the Seventh Annual BCI Research Award, with chapters that review the most promising new BCI research. As with prior years, we announced the first, second, and third place winners as part of a major international BCI conference. The Gala Awards ceremony for the 2017 BCI Research Award was part of the Seventh International BCI Conference in Graz, Austria. This conference series occurs every two years, and we already plan to host the ceremony for the 2019 award with the Eighth International BCI Conference.

Keywords BCI · Brain-computer interface · Award · EEG · ECoG

1 The 2017 Winners

As with prior years, we announced the first, second, and third place winners as part of a major international BCI conference. The Gala Awards ceremony for the 2017 BCI Research Award was part of the Seventh International BCI Conference in Graz, Austria. This conference series occurs every two years, and we already plan to host the ceremony for the 2019 award with the Eighth International BCI Conference.

Hundreds of students, doctors, professors, and other people attended the awards ceremony to see who would win. The organizer and emcee, Drs. Guger and Allison, invited a representative from each of the nominated groups to join them on the stage. All nominees received a certificate and other prizes, and remained onstage as the winners were announced. The winners for the 2017 BCI Research Award were:

C. Guger (✉)
g.tec Medical Engineering GmbH, Schiedlberg, Austria
e-mail: guger@gtec.at

B. Z. Allison
Cognitive Science Department, University of California at San Diego, La Jolla, USA

N. Mrachacz-Kersting
Department of Health Science and Technology, Aalborg University, Aalborg, Denmark

C. Guger et al. (eds.), *Brain-Computer Interface Research*,
SpringerBriefs in Electrical and Computer Engineering,
https://doi.org/10.1007/978-3-030-05668-1_11

Fig. 1 Christoph Guger (left, organizer) and S. Aliakbaryhosseinabadi (winner of the BCI Award 2017)

The 2017 BCI Award First Place Winner Is:

S. Aliakbaryhosseinabadi[1], E. N. Kamavuako[1], N. Jiang[2], D. Farina[3], N. Mrachacz-Kersting[1]

Online adaptive brain-computer interface with attention variations

1. Center for Sensory-Motor Interaction, Department of Health Science and Technology, Aalborg University, DK-9220 Aalborg, Denmark.
2. Department of Systems Design Engineering, Faculty of Engineering, University of Waterloo, Waterloo, Canada.
3. Imperial College London, London, UK (Fig. 1).

The 2017 BCI Award 2nd Place Winner Is:

Stephanie Martin[1,2], Peter Brunner[3,4], Iñaki Iturrate[1], José del R. Millán[1], Gerwin Schalk[3,4], Robert T. Knight[2,5] & Brian N. Pasley

Individual word classification during imagined speech

1. Defitech Chair in Brain-Machine Interface, Center for Neuroprosthetics, Ecole Polytechnique Fédérale de Lausanne, Switzerland
2. Helen Wills Neuroscience Institute, University of California, Berkeley, CA, USA
3. National Center for Adaptive Neurotechnologies, Wadsworth Center, New York State Department of Health, Albany, NY, USA

4. Department of Neurology, Albany Medical College, Albany, NY, USA
5. Department of Psychology, University of California, Berkeley, CA, USA

The 2017 BCI Award 3rd Place Winner Is:

Takufumi Yanagisawa[1–6]*, Ryohei Fukuma[1,3,4,7], Ben Seymour[8,9], Koichi Hosomi[1,10], Haruhiko Kishima[1], Takeshi Shimizu[1,10], Hiroshi Yokoi[11], Masayuki Hirata[1,4], Toshiki Yoshimine[1,4], Yukiyasu Kamitani[3,7,12], Youichi Saitoh[1,10]

BCI prosthetic hand to control phantom limb pain

1. Osaka University Graduate School of Medicine, Osaka, Japan.
2. Osaka University Graduate School of Medicine, Osaka, Japan.
3. ATR Computational Neuroscience Laboratories, Kyoto, Japan.
4. CiNet Computational Neuroscience Laboratories, Osaka, Japan.
5. JST PRESTO, Osaka, Japan.
6. Osaka University, Japan.
7. Nara Institute of Science and Technology, Nara, Japan.
8. University of Cambridge, UK.
9. National Institute for Information and Communications Technology, Osaka, Japan.
10. Osaka University Graduate School of Medicine, Osaka, Japan.
11. The University of Electro-Communications, Tokyo, Japan.
12. Kyoto University, Japan.

We gave the first, second, and third place winners $3000, $2000, and $1000, respectively, and other gifts. These projects explored BCIs for different goals: stroke rehabilitation, neurosurgery, and reducing phantom limb pain. One of the three winning projects involved implanted electrodes, whereas two others were non-invasive. Like earlier years, the winning projects (as well as the nominees) represent a range of experts worldwide. Among only three winning projects, the authors had affiliations from Canada and different regions within the US, EU, and Japan.

At the Award Ceremony, Dr. Guger also thanked the experts in the 2017 jury:
Natalie Mrachacz-Kersting (chair of the jury 2017),
Gaurav Sharma (Winner 2016),
Reinhold Scherer,
Jose Pons,
Femke Nijboer,
Kenji Kansaku,
Jing Jin.

2 Directions and Trends Reflected in the Awards

While the awards serve primarily to recognize the top BCI projects each year, they also provide a mechanism to explore trends within BCI research. By analysing different characteristics of projects that were submitted or nominated each year, we can see how BCI research has changed over the years. In addition to identifying new directions, we can also see which characteristics of BCI research have not changed since the first BCI Award in 2010.

For example, since 2010, the submissions and nominees have often entailed international collaboration. The groups that are most active are in the EU and US, with some contributions from other countries like China, Japan, and Canada. This international collaboration has been fairly consistent over the last several years. Similarly, projects have often involved groups with expertise in different disciplines, especially nominated projects. To develop a project that's good enough to be nominated, groups may have experts in computer science, engineering, medicine, psychology, neuroscience, and other disciplines.

Another sustained trend since 2010 has been the combination of BCIs with other devices. This combination has been quite prominent this year. The projects that were nominated in 2017 have combined BCI technology with exoskeletons, functional electrical stimulators, head-mounted displays, immersive avatars, transcranial magnetic stimulation, and other devices. BCIs may be hybridized with other devices even more often in the near future. This combination creates substantial extra challenges, but can lead to a complete system that is much more capable than a BCI by itself. Two chapters in this year's book provide good examples, showing that stimulation combined with BCI-based feedback can lead to greater benefits than a BCI without a stimulation device.

We have developed four tables to review the features of projects that were submitted more parametrically. Each table presents features of projects that were submitted

Table 1 Type of input signal for the BCI system

Property	2017% (N = 48)	2016% (N = 52)	2015% (N = 63)	2014% (N = 69)	2013% (N = 169)	2012% (N = 68)	2011% (N = 64)	2010% (N = 57)
EEG	77.1	71.2	76.1	72.5	68.0	70.6	70.3	75.4
fMRI	2.1	3.8	4.8	2.9	4.1	1.5	3.1	3.5
ECoG	8.3	11.5	9.5	13.0	9.4	13.3	4.7	3.5
NIRS	2.1	1.9	–	1.4	3.0	1.5	4.7	1.8
Spikes	4.2	7.7	4.8	8.7	7.1	10.3	12.5	–
Other signals	2.1	1.9	4.8	4.3	13.0	2.9	1.6	–
Electrodes	–	1.9	–	–	6.5	1.5	1.6	–
MEG	6.3	–	–	–	–	–	–	–
Optogenetics	2.1	–	–	–	–	–	–	–

Table 2 Submissions that used real-time versus offline signal processing

Property	2017% (N = 48)	2016% (N = 52)	2015% (N = 63)	2014% (N = 69)	2013% (N = 169)	2012% (N = 68)	2011% (N = 64)	2010% (N = 57)
Real-time BCI	95.8	94.2	96.8	87.0	92.3	94.1	95.3	65.2
Off-line applications	4.2	5.8	3.2	8.7	5.3	4.4	3.1	17.5

each year since 2010. N reflects the number of submissions, and numbers in different cells present the percentage or submissions with that feature. We present one table for each of the four general BCI components presented in the beginning of the introduction.

Sensors: Table 1 summarizes the input signals that the submitted projects used. As with previous years, we received submissions that used a wide variety of neuroimaging methods, with EEG being most prominent.

Signal processing: Table 2 shows that most submissions across the years have used on-line, real-time signal processing. Today, much of the BCI research presented in papers and at conferences includes real-time signal processing. Offline analyses often fail to capture the challenges of real-time analysis, and can present unrealistically positive results.

Output/application: Table 3 reviews the different outputs, and related applications, that have been addressed in projects submitted since 2010. There is a fair amount of change over the years, and a myriad of different applications have been explored. Consistent with other trends in the field, the 2017 submissions emphasized

Table 3 Type of output system and application

Property	2017% (N = 48)	2016% (N = 52)	2015% (N = 63)	2014% (N = 69)	2013% (N = 169)	2012% (N = 68)	2011% (N = 64)	2010% (N = 57)
Control	2.1	15.4	11.1	17.4	20.1	20.6	34.4	17.5
Platform Technology Algorithm	33.3	26.9	15.9	13.0	16.6	16.2	9.4	12.3
Stroke Neural plasticity	8.3	5.8	4.8	13.0	13.7	26.5	12.5	7
Wheelchair Robot Prosthetics	10.4	7.7	15.9	13.0	11.8	8.8	6.2	7

(continued)

Table 3 (continued)

Property	2017% (N = 48)	2016% (N = 52)	2015% (N = 63)	2014% (N = 69)	2013% (N = 169)	2012% (N = 68)	2011% (N = 64)	2010% (N = 57)
Spelling	8.3	3.8	12.7	8.7	8.3	25	12.5	19.3
Internet or VR Game	8.3	1.9	4.8	2.9	5.9	2.9	3.1	8.8
Learning	–	1.9	1.6	5.8	5.3	1.5	3.1	–
Monitoring, DOC	6.3	9.6	4.8	1.4	4.7	4.4	1.6	–
Stimulation	2.1	3.8	1.6	1.4	3.6	1.5		
Authentication Speech Assessment	4.2	3.8	4.8	13.0	3	–	9.4	–
Connectivity	–	1.9	–	–	2.4	1.5	–	–
Music, Art	–	3.8	1.6	1.4	1.8	–	–	
Sensation	–	–	–	–	1.2	–	1.6	–
Vision	–	1.9	3.2	1.4	1.2	1.5		
Epilepsy, Parkinson, Tourette's, Autism	2.1	3.8	3.2	2.9	1.2	–	–	–
Depression, Fatigue, ADHD, Pain	4.2	1.9	4.8	1.4	–	1.5	–	–
Neuromarketing, Emotion	2.1	–	–	1.4	–	1.5	–	–
Ethics	–	–	–	1.4	–	–	–	–
Mechanical Ventilation	–	–	–	–	–	–	1.6	–
Roadmap	–	1.9	–	–	–	–	–	–
Attention	4.2	–	–	–	–	–	–	–
Workload	2.1	–	–	–	–	–	–	–
Sensation	2.1	–	–	–	–	–	–	–

control less than previous years. Instead, the submissions showed a sustained or growing interest in applications to help new patient groups, such as people diagnosed with stroke or DOC and long-term amputees. Projects for mapping speech, movement, and other functions during neurosurgery have also been consistently submitted.

Environment/interaction: Table 4 presents the type of control signal that was used to interact with the BCI. Like previous years, most of the BCI systems in this year's submissions rely on one of the three types of signals that are widely used in other research. P300s and related signals were less common this year, whereas 2017

Table 4 Type of control signal used to interact with the BCI

Property	2017% (N = 48)	2016% (N = 52)	2015% (N = 63)	2014% (N = 69)	2013% (N = 169)	2012% (N = 68)	2011% (N = 64)	2010% (N = 57)
P300/N200/ERP	4.2	11.5	28.6	11.6	11.8	30.9	25	29.8
SSVEP/SSSEP/cVEP	12.5	11.5	14.3	11.6	14.2	16.2	12.5	8.9
Motor imagery	41.7	32.7	36.5	37.7	25.4	30.9	29.7	40.4
ASSR	–	–	–	–	1.8	–	1.6	–

had a higher percentage of submissions with motor imagery BCIs than any other year. This increased interest in motor imagery BCIs stems largely from BCIs for stroke rehabilitation, which we believe will become even more prominent over the next several years.

3   Interview with Dr. Martin

This year, we decided to try something new in the discussion chapter. We interviewed the lead author of the project that won second place, Stephanie Martin. She recently completed her Ph.D. under Prof. Jose Millan, one of the most well-known and respected experts in BCI research, who has both served on the 2015 BCI Research Award jury and earned two prior nominations. This is Dr. Martin's first submission, and earning a nomination—let alone second place—was a significant achievement. Congrats to Dr. Martin on finishing her Ph.D. in a top lab and her excellent project!

Decoding Inner Speech: Utopia or Reality?
Stephanie Martin studied Life Sciences and Technology at EPFL (École Polytechnique Fédérale de Lausanne), where she's been developing her skills in mathematics, physics, programming and everything that is related to the human being, including biology and biochemistry. For her master's program, she chose Neuroscience with a special focus on neuroprosthetics and neuroengineering. During her Ph.D., she dedicated her research to decoding inner speech. She submitted her work to the BCI Award 2017 and won 2nd prize. We had the chance to talk with Stephanie about her Ph.D. research and her opinion about the BCI Award.

Stephanie, how did you come up with the idea to decode inner speech?

Stephanie Martin: "In the second year of the master's program, I had the chance to go abroad. I went to UC Berkeley in California to join Robert Knight's Cognitive Neuroscience Lab where I could practice what I had learnt at the University so far. This is when I started to work on a speech decoding project. During this year abroad, I recorded brain activity of epilepsy patients who had electrodes implanted on the brain by a neurosurgeon who had to monitor their brain in order to localize seizures and epileptic foci. During the recordings, we did some experiments and language tasks with the patients. Afterwards, I analyzed the recorded ECoG data."

And then you came back to EPFL?

Stephanie Martin: "Yes. I really liked the master's project at Robert Knight's lab and I felt like I wasn't done with it. So I decided to join the Brain-Computer Interface Lab here at EPFL for my Ph.D. with the intention of continuing collaborating with Bob Knight in Berkeley, because I wanted to get the cognitive and neurological aspects as well as the engineering skills from José Millán here at EPFL, who was my professor and supervisor."

You submitted your Ph.D. project and results to the BCI Award 2017. Could you explain your Ph.D. research a little bit?

Stephanie Martin: "During my Ph.D., I continued to work on my speech decoding project, so I had to come up with new task designs with the patients and record more data. The goal was to decode inner speech and to provide an assistive technology to people who can't talk or communicate. So I was wondering if it is possible to directly decode the neural activity that was inner speech in a more natural way than other assistive technologies such as BCI, which usually allows you to control a spelling device or to move a cursor on the screen to pick one letter. I wanted to provide an alternative way of communication. It was a very difficult project because inner speech is very hard to monitor. You can't say precisely what and when people think. For instance, if I think "I am hungry", then the questions are: When did I start to think that? When did I finish thinking it? It's difficult to label the data. So, my task was to design experiments that allow me to label data or to know what and when people are thinking, and to come up with the best algorithm to extract the information I needed. That was quite a challenge that I had to face during my Ph.D."

Did you manage to decode inner speech?

Stephanie Martin: "Yes. There were several different aspects I tried to investigate during my Ph.D. because inner speech is still very unknown. Your "thinking" can be abstract. It can be a representation or it can be your own voice in your head. I tried to analyze different speech representations that are also encoded when you have your inner speech active. For instance, when you speak out loud, you can hear the sound or you have phonetic decompositions. You have words and semantics. And we know how those aspects are encoded. I investigated if those aspects are also encoded during inner speech, and if so, whether I can identify if a person is thinking one word or another."

What does this mean for Brain-Computer Interface technology?

Stephanie Martin: 'If it is possible to decode naturally whether a person is thinking 'Yes' or 'No' or 'Hungry' or 'Pain', a few clinically relevant words, then this would be the next step for future BCI. We showed that we could predict what word a person is producing internally in his/her mind. That's why we submitted my Ph.D. research to the BCI Award 2017.'

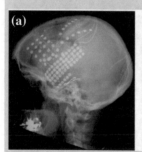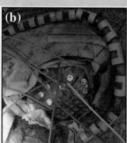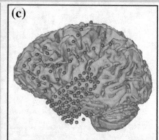

Photo These three images show an ECoG grid and surgical placement. Brain surgeons sometimes need to very precisely monitor activity over different brain areas, which requires placing ECoG electrodes on the surface of the brain. **a** Radiography of electrode placement. **b** ECoG surgical placement. **c** Electrode positions in situ

How do you imagine future BCIs in terms of language?

Stephanie Martin: "Currently, all the results we showed were obtained from people with healthy language function and have been analyzed offline. We recorded epilepsy patients in the hospital, then we got the data, then we analyzed the data, and finally, we saw the results. The next step is to go online—to replicate these results in real time, decode brain activity and thinking from people with language disabilities (e.g. aphasic patients), and then speak it out loud as a proof of concept. But I think we are very far away from this; I would say it's unrealistic in the near future. For instance, the results we showed only classified one word versus another word, a "yes" versus a "no". But there are still many steps to improve the accuracy with BCI. Now, it's really difficult to extract the information, because of the signal-to-noise ratio or the electrode location. The questions remain: How can we improve the accuracy and move to a more realistic speech device?"

What were the biggest challenges or the most successful moments for you?

Stephanie Martin: "Well, during the Ph.D. you have always your ups and downs. Research is like a rollercoaster. I think it's a big challenge to explore inner speech because, unlike when you speak out loud or when you hear speech, you can't know exactly when the brain responded and mark the data. And in addition, the brain activity is much stronger. You have beautiful data and brain activity. With inner speech, everything becomes more blurred because separating signal from noise is much harder. If I say the word "Hungry" ten times, then you have speech irregularities because brain activity is never going to be the same. In addition, you don't know when the person starts to think, the onset, the offset, so it becomes difficult to extract the information. You know more or less that inner speech happens at a certain moment, but you don't know when the different sounds happen. That is a bit of a frustrating part. There is still the question: How do you come up with an algorithm that deals with these specific issues? And how do you design tasks that can best exploit the capabilities of this new algorithm? And then, I think that's the Ph.D. life. You want to find solutions. That's when I tried to adapt classical algorithms to the specific problem. At the end, when we got the results, it was a moment of Euphoria. It was really exciting to see the outcome of so much work.""

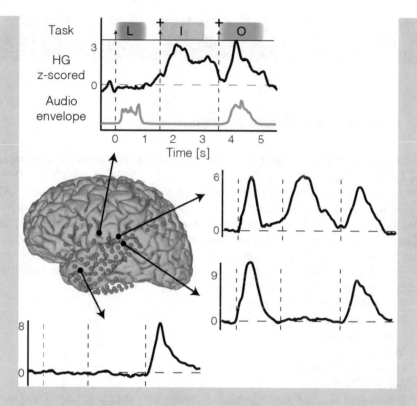

Photo Time course of changes in brain activity. Brain activity is averaged across trials and z-scored with respect to the pre-auditory stimuli baseline condition (500 ms interval) for different electrodes. The top plot displays the different conditions for the word repetition task (L = listening, I= inner speech, O = overt speech), with an example of averaged time course for a representative electrode and the averaged audio envelope (red line)

What do you think about the International BCI Award?

Stephanie Martin: "Originally, I thought that my research was a bad candidate for the BCI Award because I thought it has to be a closed-loop Brain-Computer Interface, something that works, something that shows an interface. After discussing this with my supervisor José Millán, who suggested that my speech decoding research was a good candidate, even if there is not a BCI per se included, I submitted my project for the 2017 BCI Award. I thought that my project might be relevant for the field of BCI. It could open doors to new BCI applications. The BCI Award is open to any projects that are not necessarily BCIs per se, but relevant for the field. I won the 2nd Prize in the BCI Award 2017, which was really encouraging. And in the future, I think I will move to a more closed-loop BCI to increase our chances and maybe win 1st place. The BCI Award is a top prize in the BCI field and renowned internationally. I was happy to include it in my CV and hope to add another BCI Award someday!"

Thanks Stephanie!

4   Conclusion and Future Directions

Each year, we produced a book that reviews the projects nominated for the Annual BCI Research Award and considers trends that are reflected in the awards. These books and awards have helped to encourage and disseminate top-quality BCI research, and we are considering new options for awards and book chapters. For example, we may expand the awards by adding different categories, and we're considering adding a chapter focused more on interviews with one or more BCI research groups about their project and experiences.

But, our main focus is on the annual awards. The 2018 Award ceremony occurred fairly early, in May 2018. The 2018 jury was:

Kai Miller (chair of the jury 2018),
Natalie Mrachacz-Kersting (winner 2017),
Vivek Prabhakaran,
Yijun Wang,
Milena Korostenskaja,
Sharlene Flesher.

Over half of the 2018 jury consisted of experts who were nominated for at least one BCI Research Award in 2017 or earlier. The chair of the jury, Kai Miller, was nominated in 2011 and 2014. The jury included two nominees from the 2017 Awards, Natalie Mrachacz-Kersting and Milena Korostenskaja. Sharlene Flesher was nominated for two different projects in 2016. While the jury typically included the winner from the preceding year (in this case, Prof. Mrachacz-Kersting), we were fortunate this year to have three other jury members who know what it's like to develop a project that is good enough to be nominated.

This eclectic jury has expertise in medicine, physics, neuroimaging, signal processing, rehabilitation, engineering, and other disciplines that are important in BCI research. They have worked with systems from different companies and work at institutes across the EU, US, and Japan. They include people who work with both implanted and non-invasive neuroimaging methods, from both medical and research sectors. This is the kind of jury that is necessary to evaluate the newest projects in BCI research, which often involve complex collaborations across disciplines, systems, regions, and sectors.

To conclude, the 2017 BCI Awards and the resulting chapters have introduced and recognized many of the most innovative and promising new projects in the BCI research community. Most of the nominees come from well-known, established groups that are currently exploring even newer directions based on their nominated projects. We have also explored different trends in BCI research by analysing different characteristics of the submissions. The Annual BCI Research Awards and books will continue, and nominees from the 2018 BCI Research Awards have already begun writing their chapters.

Printed in the United States
By Bookmasters